Series:

Computational Product Design and Digital Manufacturing

Series editor: Prof. Panagiotis Kyratsis

Department of Product and Systems Design Engineering
University of Western, University Campus Kila
Kozani, Greece

Series: Computational Product Design and Digital Manufacturing

Hybrid Materials and Advanced Manufacturing

Techniques and Applications, Volume I

Editors:

Rajesh Kumar Verma
Professor, Department of Mechanical Engineering
Harcourt Butler Technical University, Kanpur, India

Devendra Kumar Singh
Senior Research Assistant in Department of Mechanical Engineering
Harcourt Butler Technical University, Kanpur, India

Arpan Kumar Mondal
Assistant Professor, Department of Mechanical Engineering
National Institute of Technical Teachers' Training & Research
Kolkata, India

Panagiotis Kyratsis
Professor, Department of Product and Systems Design Engineering
University of Western, University Campus Kila Kozani, Greece

CRC Press
Taylor & Francis Group
Boca Raton London New York

CRC Press is an imprint of the
Taylor & Francis Group, an **informa** business
A SCIENCE PUBLISHERS BOOK

First edition published 2026
by CRC Press
2385 NW Executive Center Drive, Suite 320, Boca Raton FL 33431

and by CRC Press
4 Park Square, Milton Park, Abingdon, Oxon, OX14 4RN

CRC Press is an imprint of Taylor & Francis Group, LLC

Library of Congress Cataloging-in-Publication Data (applied for)

ISBN: 978-1-032-99906-7 (hbk)
ISBN: 978-1-041-00035-8 (pbk)
ISBN: 978-1-003-60735-9 (ebk)

DOI: 10.1201/9781003607359

Typeset in Times New Roman
by Prime Publishing Services

Preface

Rapid advances in hybrid materials and manufacturing technologies are reshaping modern engineering, driving innovation in aerospace, biomedicine, automotive, and sustainable energy. Hybrid Materials and Advanced Manufacturing: Modern Techniques and Applications (Volume I) offers a comprehensive look at the intersection of material innovation and advanced fabrication. This volume opens with foundational insights into composite materials—their microstructure, physio-mechanical traits, and industrial applications. It then explores innovations like green-synthesized hybrid nanoparticles, alginate-based biocomposites for remediation, and additive manufacturing. Emphasis is placed on laser cladding, automation, and strategies to enhance composite performance, ensure structural integrity, and support cost-effective, defect-free sustainable production.

This volume connects theory with practice in areas like flexible robotics and circular economy, blending sustainability with innovation. Aimed at students, researchers, and professionals, it serves as both a learning tool and a springboard for next-gen solutions in hybrid materials and advanced manufacturing.

Acknowledgements

The completion of this volume was made possible through the collective expertise and unwavering support of various individuals and institutions. We are especially grateful to CRC Press, Taylor & Francis Group, for their guidance, and to the Uttar Pradesh Council of Science and Technology (UPCST), Lucknow, under the Department of Science and Technology, Government of Uttar Pradesh, whose R&D Project (ID: CST/D 1267) significantly strengthened our research infrastructure and enabled key advancements in hybrid materials and advanced manufacturing. We sincerely thank the Department of Science and Technology (DST), Government of India, for their vital support under the TIDF-IITG scheme (Project ID: 1019). We also extend heartfelt appreciation to Prof. Samsher, Hon'ble Vice Chancellor of HBTU Kanpur, for encouraging interdisciplinary collaboration. Our deep gratitude goes to all contributing authors and global collaborators whose dedication was key to the success of this project.

We sincerely thank Dr. Jogendra Kumar and Mr. Virat Mani Vidhyasagar (HBTU Kanpur), along with Dr. Shivi Kesarwani and Dr. Kuldeep Kumar (MMMTU Gorakhpur), for their dedicated support in this project. Our gratitude also extends to the reviewers, editorial board, and CRC Press, Taylor & Francis team for their invaluable contributions in shaping this book—this achievement would not have been possible without their consistent guidance and collaboration.

Contents

List of Contributors

Abhishek Verma
Department of Polymer and Process Engineering, Indian Institute of Technology Roorkee, Saharanpur Campus, India

Akshay Kumar Shakya
Mechanical Engineering, Sanskar College of Engineering & Technology, Ghaziabad, India

Alok Kumar
Department of Mechanical Engineering, Indian Institute of Technology, Jodhpur, India

Alpay Tamer Ertürk
Mechanical Engineering Department, Kocaeli University, Umuttepe Campus, 41001, Izmit, Kocaeli, Turkiye

Amit Kumar
Indian Institute of Technology, Ropar, India

Anjali Awasthi
Department of Chemical Engineering, Harcourt Butler Technical University, Kanpur, India

Ashish Kapoor
Department of Chemical Engineering, Harcourt Butler Technical University, Kanpur, India

Chandrmani Yadav
Marwadi University Research Center, Faculty of Engineering & Technology, Department of Mechanical Engineering, Marwadi University, Rajkot, Gujarat, India

Darshan Srivastav
Department of Applied Science & Humanities, Maharana Pratap Institute of Technology, Gorakhpur, India

Deepansh Kumar
Darbhanga College of Engineering, Department of Science Technology and Technical Education, Darbhanga, India

Dipti Bharti
Darbhanga College of Engineering, Department of Science Technology and Technical Education Darbhanga, India

Disha Mishra
Department of Applied Science & Humanities, Maharana Pratap Institute of Technology, Gorakhpur, India

Harish Sharma
Department of Paper Technology, Indian Institute of Technology Roorkee, Saharanpur Campus, India

Hemalata Jena
School of Mechanical Engineering, KIIT Deemed to be University, Bhubaneswar 751024, India

Hina Fatina
Department of Polymer and Process Engineering, Indian Institute of

Technology Roorkee, Saharanpur Campus, India

Manas Paramanik
Epidemiology, Vector Biology and Environmental Monitoring Research Units, Entomology Laboratory, Department of Animal Science, Kazi Nazrul University, Asansol, India

Manoj Kumar
Department of Mechanical Engineering, RTC Institute of Technology, Ormanjhi, India

Meghanshu Vashista
Department of Mechanical Engineering, Indian Institute of Technology (BHU) Varanasi, Varanasi, India

Mohd Zaheer Khan Yusufzai
Department of Mechanical Engineering, Indian Institute of Technology (BHU) Varanasi, Varanasi, India

Nathi Ram Chauhan
Department of Mechanical & Automation Engineering, Indira Gandhi Delhi Technical University for Women, Delhi, India

Neslihan Başaran
Department of Materials Science and Engineering, Gebze Technical University, Gebze, Turkey

Neslihan Tamsü Selli
Department of Materials Science and Engineering, Gebze Technical University, Gebze, Turkey

Nishu Sharma
Department of Mechanical & Automation Engineering, Indira Gandhi Delhi Technical University for Women, Delhi and Assistant Professor, Mechanical Engineering, Mangalmay Group of Institutions

Pranava Chaudhari
Department of Chemical Engineering, Harcourt Butler Technical University, Kanpur, Uttar Pradesh, India

Prashant Singh
Department of Polymer and Process Engineering, Indian Institute of Technology Roorkee, Saharanpur Campus, India

Priya Singh
Department of Chemistry, Harcourt Butler Technical University Kanpur, India

Rahul Singh
School of Basic and Applied Sciences, IILM University, Greater Noida, India

Ravi Prakash Ojha
Department of Chemistry, Shri Lal Bahadur Shastri Degree College, Gonda, India

Ravi Shankar Rai
Department of Automation and Robotics Engineering, JSPM's Rajarshi Shahu College of Engineering, Pune, India

Rohit Kumar
Darbhanga College of Engineering, Department of Science Technology and Technical Education, Darbhanga, India

Sagarika Mukherjee
Epidemiology, Vector Biology and Environmental Monitoring Research Units, Entomology Laboratory, Department of Animal Science, Kazi Nazrul University, Asansol, India

Sandeep Sagar
Darbhanga College of Engineering, Department of Science Technology and Technical Education, Government of Bihar, Darbhanga, India

Sanjeev Kumar Sharma
Department of Applied Science & Humanities, Maharana Pratap Institute of Technology, Gorakhpur, India

Santosh Kumar
Department of Chemistry, Harcourt Butler Technical University Kanpur, India

Sudhir Agrawal
Department of Computer Science and Engineering, Maharana Pratap Institute of Technology, Gorakhpur, India

Sudip Paramanik
Epidemiology, Vector Biology and Environmental Monitoring Research Units, Entomology Laboratory, Department of Animal Science, Kazi Nazrul University, Asansol, India

Suman Dasmodak
Epidemiology, Vector Biology and Environmental Monitoring Research Units, Entomology Laboratory, Department of Animal Science, Kazi Nazrul University, Asansol, India

Sunil kumar Yadav
Department of Mechanical Engineering, Madan Mohan Malaviya University of Technology, Gorakhpur, India

Vikas Diwakar
Department of Mechanical Engineering, Indian Institute of Technology (BHU) Varanasi, Varanasi, India

Vinod Baniya
Department of Applied Science & Humanities, Maharana Pratap Institute of Technology, Gorakhpur, India

CHAPTER

1 Fundamentals of Composite Materials

Abhishek Verma,[1,*] *Prashant Singh,*[1] *Hina Fatina*[1] and *Harish Sharma*[2]

Composite materials are engineered by combining two or more distinct substances to enhance thermal, mechanical, or electrical properties. They consist of a matrix (polymer, metal, or ceramic) that binds and protects, and a reinforcement (fibers or particles) that adds strength and stiffness. By combining lightweight, flexible polymers with strong carbon fibers, composites excel in aerospace and automotive use. Performance depends on fiber length, reinforcement content, and interfacial bonding. Their key benefits include high strength, low weight, corrosion resistance, flexibility, and design versatility. Efficient composites face challenges like recycling issues, high production costs, and delamination. Research focuses on sustainable materials, advanced manufacturing, and predictive modeling. Their strength and versatility make them vital in aerospace, energy, construction, water treatment, and safety. Environmental concerns, especially recyclability and sustainability, remain key topics.

1.1 Introduction to Composite Materials

Composite materials have transformed engineering by solving complex material challenges through the combination of distinct constituents (Ramesh et al.,

[1] Department of Polymer and Process Engineering, Indian Institute of Technology Roorkee, Saharanpur Campus, India.
[2] Department of Paper Technology, Indian Institute of Technology Roorkee, Saharanpur Campus, India.
* Corresponding author: abhishek_v@pe.iitr.ac.in

DOI: 10.1201/9781003607359-1

2025). This synergy yields superior properties unattainable by metals, ceramics, or polymers alone (Bhong et al., 2023). Natural examples like wood (cellulose in lignin) and bone (collagen in mineral matrix) illustrate the effectiveness of composite structures (Morwenna Spear, 2018). Modern composite development began in the mid-20th century with fiberglass and carbon fiber-reinforced polymers (Sivasubramanian et al., 2021). Today, composites are widely used in aerospace, automotive, sports, biomedical, and packaging sectors. They consist of a matrix and reinforcement: the matrix binds and protects, while the reinforcement, facilitating the effective transfer of loads between the two components. The reinforcement in the form of flakes, fibers, or particles reinforce strength and stiffness. Together, they form lightweight, strong, and highly customizable materials (Diniță et al., 2023). Composite science explores how material structure, interfaces, and surface treatments affect performance. Composites typically begin as petroleum-based raw materials, processed into monomers, polymerized, and combined with reinforcements. Their applications vary by properties, and after use, they may be recycled, reused, incinerated for energy, or landfilled. Figure 1.1 illustrates the complete life cycle of plastic composite materials.

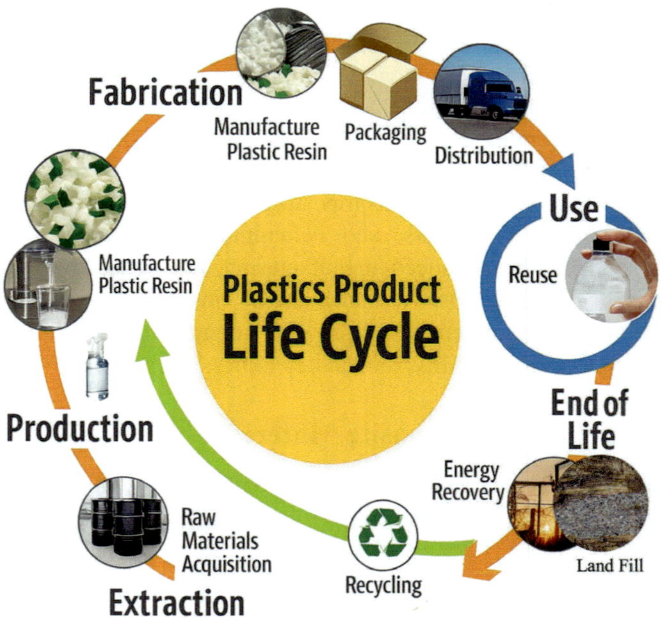

Figure 1.1: Life cycle of the plastic composite.

Composite research and application demand thorough testing, characterization, and quality control. Stability depends on material properties, composition, processing, testing methods, and time effects. This chapter highlights standardized, systematic testing across structural levels to ensure reliable composite performance.

1.2 Key Components of Composite Materials

1.2.1 The Matrix

The matrix is the continuous phase in a composite that surrounds and supports the reinforcement, influencing properties like durability, thermal stability, and chemical resistance. It typically falls into one of three categories: polymer, metal, or ceramic.

(a) **Polymer Matrix:** Polymer matrices are widely used for their lightweight, flexibility, and easy processing. Common examples include epoxy, vinyl ester, and polyester resins. They are prevalent in automotive parts, wind turbine blades, and sports gear.

(b) **Metal Matrix:** Metal matrix composites (MMCs), using metals like aluminum, magnesium, or titanium, offer high strength and thermal conductivity for demanding environments. They are commonly applied in automotive and aerospace components.

(c) **Ceramic Matrix:** Ceramic matrix composites (CMCs), made from materials like silicon carbide and alumina, are designed for high-temperature, wear-resistant applications such as jet engines, brake systems, and nuclear reactors.

Poor bonding between fibers and the matrix can lead to microcracks, delamination, and failure. Composite usability is also affected by matrix viscosity. Common matrix materials are polymers, mainly classified as thermosets or thermoplastics (Wu et al., 2023; Kargarzadeh et al., 2017; Hasan et al., 2019). Figure 1.2 and Table 1.1 illustrate their classifications and usage in composite fabrication.

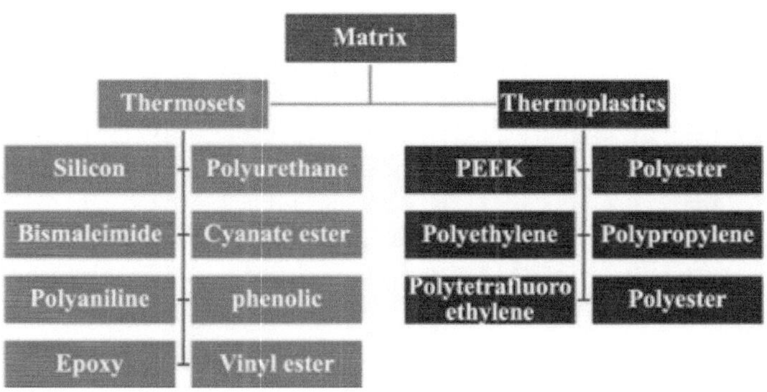

Figure 1.2: Various matrix types used in composites, categorized into thermosets and thermoplastics (Muthalagu et al., 2021).

Table 1.1: Various materials used as the matrix in the preparation of composites.

No.	Fiber material	Matrix material	Achievement	Reference
1	Glass	Epoxy	Water-saturated graphene-oxide/glass-epoxy composites show enhanced mechanical properties.	Muthalagu et al., 2021
2		Polytetrafluoroethylene	Steel mesh-reinforced GFRP improves wear resistance.	Zhu et al., 2019
3		Polyester	Void volume fraction critically affects mechanical performance of GFRP.	Demircioglu et al., 2021
4		Polyimide	Porosity and voids significantly influence heat transfer in space environments.	Pilato and Michael Michno, 1994
5		Epoxy	Reduced interfacial thermal resistance enhances thermal conductivity.	Alves et al., 2020
6		Bismaleimide	Aged composites absorb electromagnetic radiation efficiently.	Hasan et al., 2019
7		PEEK	Surface treatments improve composite mechanical properties.	Sano et al., 2018
8		PEEK	Flame exposure duration slightly impacts tensile strength.	Patel and Chaudhary, 2022
9	Carbon	Polyethylene	Carbon black-loaded composites outperform those with activated carbon black.	Plastics Europre, 2025
10		Polyamide	Polyamide composites of polyamides aid in thermal regulation systems.	Plastics Europre, 2025
11		Epoxy	Epoxy-carbon composites reduce friction at high temperatures.	Pascault and Williams, 2018
12		Polyurethane	Hydrogen bonding boosts thermal properties in polyurethane/CNT composites.	Anand et al., 2021
13		Polyaniline	Pt-Ru/polyaniline carbon composites serve as efficient methanol oxidation catalysts.	Bandaru et al., 2021
14		Bismaleimide	Diaminodiphenyl methane used in carbon fiber reinforced bismaleimide composites.	Sim et al., 2021
15		Bismaleimide	Nitrogen-doped carbon obtained by degrading bismaleimide resin.	Liu et al., 2021

16		Cyanate ester	Copper incorporation enhances electrical conductivity.	Ren et al., 2021
17		Cyanate ester	Addition of manganese dioxide and polyamide-6 increases tensile strength by 98%.	Xue et al., 2019
18		Cyanate ester	Atomic oxygen exposure boosts interlaminar shear strength by approximately 43%.	Pan et al., 2021
19		PEEK	Graphene oxide coatings on show promise in biomedical fields.	Vieille et al., 2022
20	Aramid	Polypropylene	Tensile strength predicted using modified rule of mixtures and the void content.	Choudhury et al., 2021
21		Polyimide	Weight reduction achieved through thermoset polyimide incorporation.	Xu et al., 2021
22		Polypropylene	Branched fibers and transcrystalline layers enhance interlocking and shear strength.	Liang et al., 2021
23		Epoxy	CFRB-AFRB sandwich panel design enhances energy absorption.	Li et al., 2021
24		Epoxy	MWCNTs improve the mechanical properties of AFRP composites.	Zhang et al., 2021

1.2.2 The Reinforcement

The reinforcement in a composite is the discontinuous phase that imparts strength and hardness to the matrix. Based on geometry, reinforcements are classified into three types: fibers, particles, and flakes.

(a) **Fiber Reinforcements:** Fiber reinforcements—such as glass, carbon, and aramid fibers—are the most widely used in composites, enhancing strength and stiffness along their length. They're essential in aerospace, automotive, sports, and construction applications.

(b) **Particle Reinforcements:** Particle reinforcements like alumina and silicon carbide enhance composite hardness, wear resistance, and thermal conductivity. They're widely used in wear-resistant coatings and cutting tools.

(c) **Flake Reinforcements:** Flake reinforcements, such as mica and graphene, improve barrier and electrical properties in composites, making them ideal for packaging and electronics.

1.3 Characterization of Composite Materials

Multiple techniques assess composite properties and processing performance. TGA, IR, and GPC are used for material characterization (Shrikaant Kulkarni, 2019). DSC and TG evaluate resin cure temperature, enthalpy and Tg, while TG also reveals volatile and fiber content in prepregs (Lo and Nutt, 2019). DMA traces modulus, softening, gel, and hardening points to optimize cure profiles (Stark, 2013). GPC profiles molecular weight distribution in prepregs, indicating aging resistance via polymer/monomer ratios (Smith and Hubert, 2023). Since matrix and fiber content may vary during storage or transport, routine characterization ensures composite quality. Key composite properties—fiber volume, ply thickness, void content, Tg, moisture uptake, and dimensional stability—are vital for quality control. Tensile tests, using straight-sided, tabbed, untagged, or dog-bone specimens, assess strength, modulus, and fracture strain. Tabbed and untagged designs suit specific fiber orientations, while dog-bone shapes require precise machining. Compression tests employ long/short gauge specimens or honeycomb cores; short gauge is more reliable under varied temperatures. Shear tests use ±45°-off-axis tension or double V-notched specimens, the latter ensuring uniform stress fields (Ku et al., 2021; Haluza et al., 2022).

Certain tests—like rail-shear, thin-wall torsion, and off-axis tension—can yield errors due to edge effects and non-uniform stress (Pecić et al., 2023). Interlaminar strength is typically assessed by Short Beam Shear (SBS), though it doesn't provide pure shear values. Bending tests often undervalue modulus but approach tensile values as span-to-thickness ratio increases; a ratio of 40 is preferred for accuracy. Other mechanical evaluations include quasi-static indentation, compression after impact, and mode II interlaminar fracture toughness. ASTM D6856-03 outlines testing standards for fabric textile composites, while ASTM D4672-04 details methods for continuous fiber-reinforced polymer composites. ASTM D149-97a, D150-98, D495-99, and D2303-97 address laminate and lamina testing. Key methods for evaluating shear strength, stress-strain, and modulus include V-notched specimens, rail-shear, hoop-wound cylinders, and out-of-plane tensile tests (e.g., interlaminar and curved laminate strength), each with distinct pros and cons (Carey et al., 2024).

1.4 Design Parameters and Manufacturing of Composites

The design and fabrication of composite materials require a thorough understanding of matrix–reinforcement interactions. Key factors influencing composite performance include:

(a) Fiber Orientation: Fiber orientation governs composite mechanics—unidirectional fibers yield high directional strength, while random orientation ensures more isotropic properties.

(b) Volume Fraction: The ratio of reinforcement in a composite—directly impacts strength, stiffness, and density. Higher fractions typically enhance mechanical performance.

(c) Interfacial Bonding: Interfacial bonding governs load transfer between matrix and reinforcement, critically affecting performance. Poor bonding may cause delamination and structural failure. Testing of fiber, resin, and prepreg variations ensures compliance with material and processing standards. Alam et al. (2014) detail advanced materials in civil aircraft, focusing on carbon fiber composites and *ex-situ* methods to enhance impact resistance and energy absorption (Alam et al., 2014). They also highlight the promise of conductive polymer composites in aerospace applications.

1.4.1 Manufacturing of Composites

Preparation techniques for polymeric composites depend on matrix and fiber types, production volume, and cost-effectiveness. Manufacturing is key to tailoring structural properties (Neher et al., 2014). Initially developed via hand lay-up, fiber-reinforced composites now utilize methods like filament winding, RTM, compression molding, and extrusion. Plastic industry processes (e.g., injection molding, calendaring) are adapted for thermoplastics. Molding enables complex shapes and enhances performance and durability, with composite type and application guiding method selection.

Table 1.2: Summary table of the manufacturing techniques.

Manufacturing Technique	Advantages	Limitations	Typical Applications	Cost Implications	Reference
Hand Lay-Up	Simple, low-cost tooling, ideal for prototyping	Labor-intensive, inconsistent quality	Boat hulls, automotive panels	Low initial cost; high labor cost	Middleton, 2016
Bag Molding (Vacuum Bagging)	High fiber content, improved laminate quality	Requires autoclave or vacuum equipment	Aerospace components, structural laminates	Moderate to high	Vita et al., 2019
Filament Winding	Excellent strength in hoop direction, automation-friendly	Limited to axisymmetric parts	Pressure vessels, tanks, pipes	Moderate	Lee and Springer, 1990; ASM International, 2011
Pultrusion	Continuous production, high fiber volume, low labor	Restricted to constant cross-sectional shapes	Rods, beams, structural profiles	Cost-effective for large-scale production	Moschiar et al., 1998
Plasma Spraying	High-temperature coatings, good adhesion	Complex setup, high energy consumption	Biomedical coatings, heat shields	High	Cheang and Khor, 1996
Magnetron Sputtering	Thin uniform coatings, precise control	Slow process, expensive equipment	Electronics, advanced composites	High	Neelesh K. Jain et al., 2016
Electrochemical Deposition	Uniform coatings, low temperature process	Limited material selection, surface prep required	Biomedical implants, catalysts	Moderate	Jain et al., 2016
Sol-Gel Technology	Low-temp processing, high purity coatings	Time-consuming, moisture-sensitive	Optical coatings, barrier films	Moderate	Yamaguchi et al., 2006
Resin Transfer Molding (RTM)	Good surface finish, suitable for complex shapes	Moderate tooling cost, control needed for resin flow	Automotive parts, large structures	Moderate to high	Fong and Advani, 1998
Compression Molding	Suitable for high-strength thermosets, short cycle time	High mold cost, material waste	Automotive components, appliance housings	High initial tooling, low unit cost	Jaafar et al., 2019
Injection Molding	High production rate, precise and repeatable	Not ideal for long fibers or large parts	Small, detailed thermoplastic parts	High setup cost, low per-unit cost	Fernandes et al., 2018
Extrusion	Continuous production, efficient for large volumes	Limited to uniform cross-sections	Pipes, sheets, films	Low for thermoplastics	Repka et al., 2012

1.4.1.1 Hand Lay-Up

Hand lay-up is the most widely used method for fabricating fibrous-reinforced composites. It involves manually layering glass-fiber mats or fabrics with polyester or epoxy resins onto open molds (wood, metal, or reinforced plastic). A release agent ensures easy demolding, and a gel coat is applied for surface finish. Air is removed using a roller, and in the spray-up variant, chopped fibers and resin are concurrently deposited via a spray gun (Middleton, 2016).

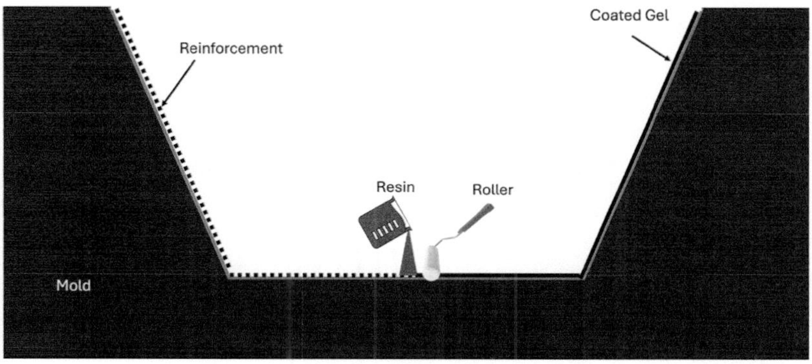

Figure 1.3: Hand Lay-Up process.

1.4.1.2 Bag Molding

Bag molding is key for producing high-quality composite parts using prepregs—fiber layers pre-impregnated with ~50% epoxy resin and stored cold to limit reactivity. The stack is vacuum-bagged and autoclaved under heat and pressure, consolidating it into a dense laminate (Vita et al., 2019). Excess resin expelled during curing helps achieve a ~60% fiber volume fraction. For more, see Bader and Dillon et al. (1997), Lekakou (1997), and Sidwell (1998).

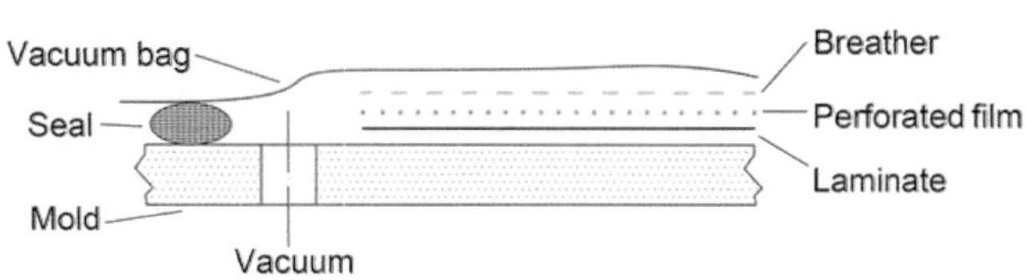

Figure 1.4: Bag molding process.

1.4.1.3 Filament Winding

Filament winding involves wrapping continuous fibers, pre-wetted with resin, onto a mandrel at controlled angles (Fig. 1.5). After curing in a high-temperature oven, the part is removed from the mandrel. This method is ideal for cylindrical or conical components like tanks and pipes. Mandrel removal can be challenging,

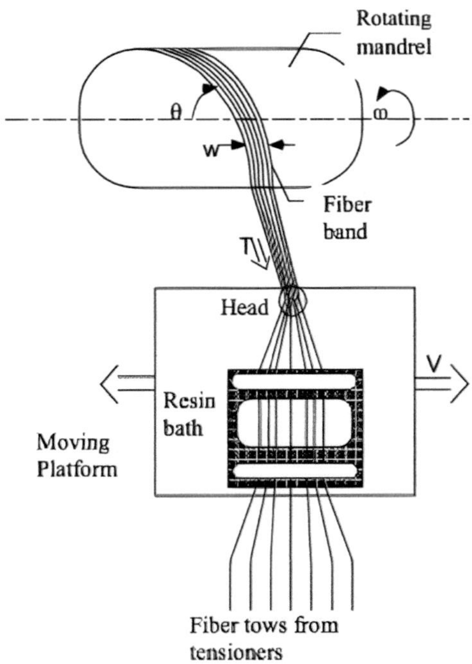

Figure 1.5: Filament winding process.

requiring solutions like inflatable, segmented, or dissolvable mandrels (Lee and Springer, 1990; ASM International, 2011).

1.4.1.4 Pultrusion

Pultrusion is a continuous, automated process for fabricating constant cross-section profiles like rods and tubes. Glass fiber mats pass sequentially through a resin bath, guide system, forming die, and heated die, triggering rapid matrix polymerization (Fig. 1.6). The composite is then cooled and cut by a flying saw without interrupting the pulling motion. This method efficiently produces structural shapes with fiber content up to 65–76% by volume (Moschiar et al., 1998).

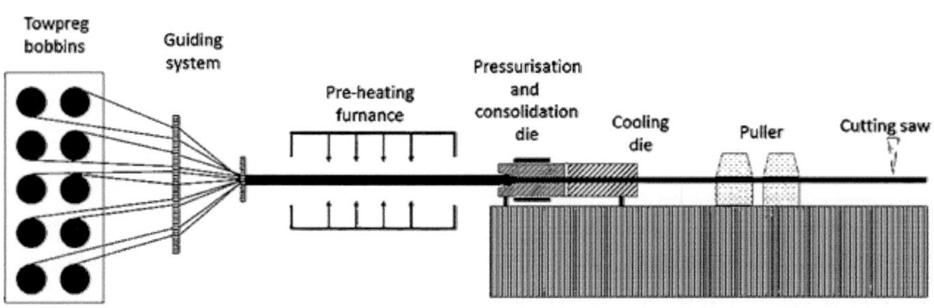

Figure 1.6: Pultrusion process.

1.4.1.5 Surface Coating

Surface coating technology enhances material performance by applying thin films to polymer or non-polymer surfaces. Common methods include plasma spraying, electrochemical deposition, magnetron sputtering, and sol-gel processes.

- **Plasma Spraying:** Plasma spraying, common in biomedicine, uses a plasma arc to heat powder to molten or semi-molten states, which is then sprayed at high velocity to form a strong coating (Fig. 1.7). This enhances heat, corrosion, and wear resistance. The coating bonds via chemical/physical adsorption and mechanical interlocking. Increased substrate roughness improves adhesion by trapping molten particles and promoting chemical bonding (Cheang and Khor, 1996).

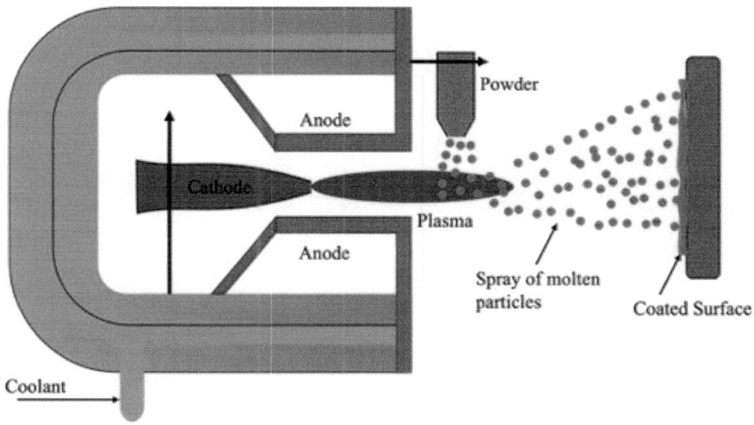

Figure 1.7: Plasma spraying process (Devasia et al., 2021).

- **Magnetron Sputtering:** Magnetron sputtering, illustrated in Fig. 1.8, uses magnetic fields to direct incident particles toward a target, causing collisions that eject surface atoms, which then deposit onto a substrate. This cascade-driven process enables thin-film formation, though the exact bonding mechanisms and film growth dynamics remain not fully understood (Jain et al., 2016).
- **Electrochemical Deposition:** Electrochemical deposition (or electroplating) involves the migration of charged particles between electrodes under an applied voltage, forming a uniform coating on a conductive substrate. It is more cost-effective and energy-efficient than plasma spraying or magnetron sputtering, operating at ambient temperatures with colloid-based systems. Widely used in biomedicine—for tissue engineering and drug delivery—it lacks a unified theoretical model, despite various experimentally inferred mechanisms (Jain et al., 2016).

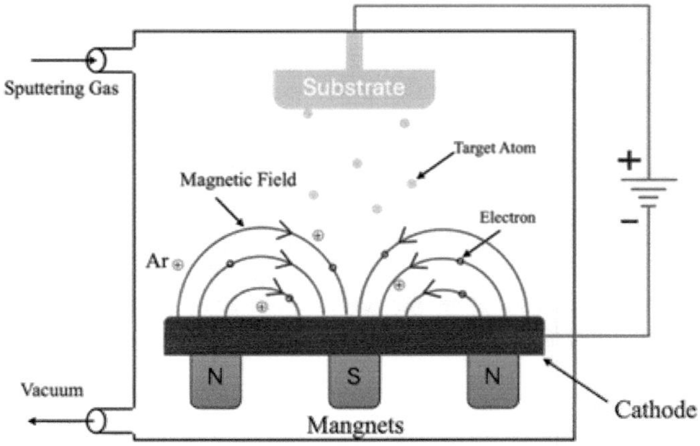

Figure 1.8: Working principle of magnetron sputtering (Jain et al., 2016).

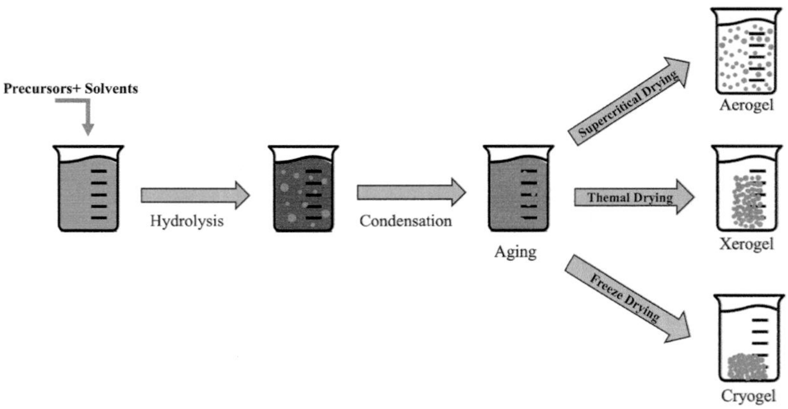

Figure 1.9: Sol-Gel technology for composite preparation (Bokov et al., 2021).

- **Sol-Gel Technology:** Sol-gel technology, introduced by Yamane et al. (1970), is commonly used to fabricate thin oxide films (Fig. 1.9) (Sakka, 2016). Figure 1.9 Substrate pretreatment is crucial, especially for polymers with poor adhesion. Techniques like ionic functionalization and chemical abrasion enhance surface properties. Yamaguchi et al., demonstrated effective alumina film deposition on PET, PMMA, and PC, achieving light reflectance values of 1.5%, 1.0%, and 0.8%, respectively. These results highlight the significant impact of substrate material on film performance (Pilato and Michno, 1994).

1.4.1.6 Composite Melt Processing

Composite melt processing began with John W. Hyatt's 1868 method of using heat to mold softened polymers. Later innovations, like screw integration, enabled simultaneous polymer mixing and recycling. *Ex situ* liquid molding further advanced the process by decoupling resin flow from reaction, allowing longer flow paths and improved layer interfaces. This enhances microstructure, boosting shear strength and impact resistance—ideal for cost-effective, high-performance materials. Integrating design, materials, manufacturing, and certification is essential for reducing costs and enhancing aerospace composite performance.

- **Resin Transfer Molding (RTM):** Resin Transfer Molding (RTM) involves placing a fibrous preform in a closed mold, then injecting resin under pressure. The mold's cure cycle is tailored to the resin system and part thickness to ensure adequate green strength for demolding (Fig. 1.10). Final curing is completed in an electric oven. RTM offers a cost- and time-efficient method for producing complex, near-net shapes versus traditional lay-ups. For further details, see Fong and Advani (1998).

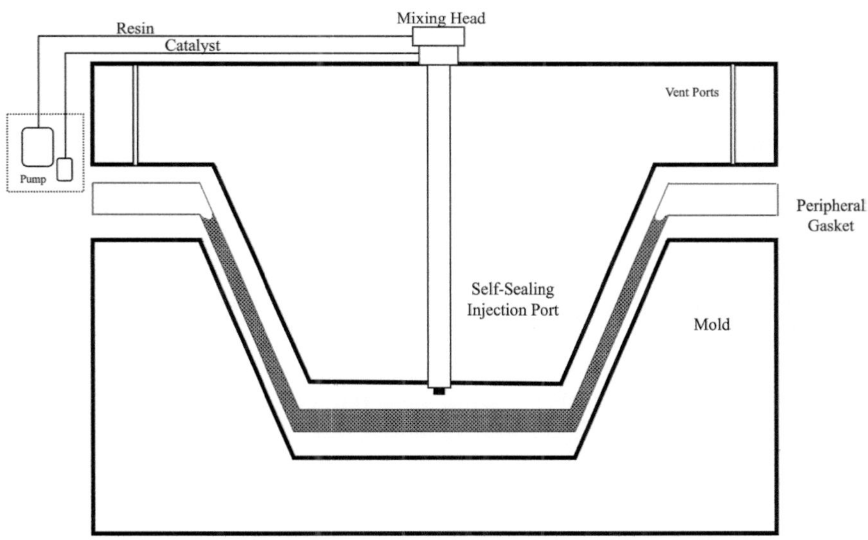

Figure 1.10: Resin transfer molding process (Fong and Advani, 1998).

- **Compression Molding:** Compression molding—illustrated in Figure 1.11—is an early mass-production method for polymer products. It involves placing preheated material into an open, heated mold, which is sealed using a top force to ensure full surface contact. Under heat and pressure, the polymer cures. Thermoplastics require post-molding cooling under pressure, making the process impractical for continuous runs. In contrast, thermosets cure through cross-linking and can be ejected without cooling, though achieving full mold filling can be challenging (Jaafar et al., 2019).

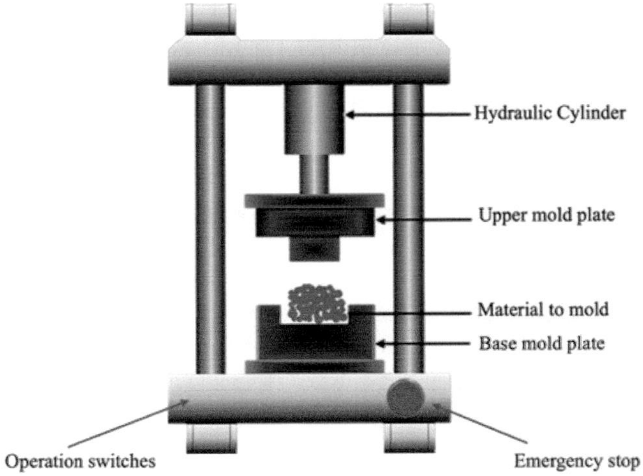

Figure 1.11: Compression molding.

- **Injection Molding:** Injection molding is a solid-state process ideal for forming complex shapes with smooth surfaces. **Heated polymer** granules or short fiber mixtures are injected into a mold under air pressure (Fig. 1.12). The "50% rule" recommends retaining half the material inside a heated cylinder for consistent mixing and residence time; 80% may be used for sensitive materials, though it increases degradation risk. In nanocomposites, nanoparticles are pre-dispersed using masterbatches or extrusion. Process control—temperature, pressure, injection speed—is critical for reproducibility. According to Yousefian and Rodrigue (2015) (Kargarzadeh et al., 2017), mold temperature affects crystallization, while injection temperature influences viscosity and particle alignment. High processing temperatures may degrade cellulose nanoparticles (Fernandes et al., 2018).

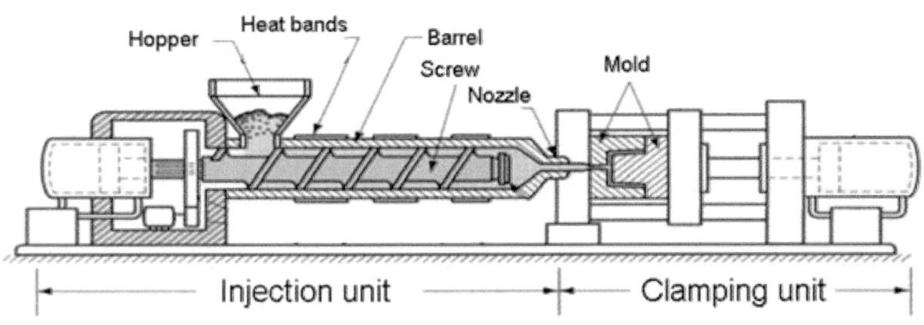

Figure 1.12: Injection molding process.

- **Extrusion:** Extrusion and injection molding enable rapid, reliable fabrication of polymer composites. Melt processing is preferred over solvent casting for cellulose nanocomposites due to scalability. In extrusion, screw-induced stress disperses particles, with rotation speed affecting mechanical properties (Nicharat et al., 2015). Low-shear mixing better preserves particle structure. Short retention times and screw action hinder percolation networks. During molding, stress aligns rod-like particles—alignment may persist after solidification but can partially reverse (Repka et al., 2012).

1.5 Mechanical Failure in Composite Materials

Despite their impressive strength and flexibility, composites face limitations—most notably mechanical failure, which threatens long-term reliability. The three most common failure modes are delamination, fatigue, and matrix cracking (Etri et al., 2024).

1.5.1 Delamination

Delamination is a failure mode in laminated composites where layers separate during operation. Common causes include poor fiber–matrix bonding, impact loads, manufacturing defects (e.g., voids or trapped air), and environmental factors such as moisture ingress and thermal cycling. Typically resulting from repeated loading, delamination reduces structural load-bearing capacity. Its concealed nature makes detection difficult, and surface inspection is generally ineffective. This degradation significantly lowers interlaminar shear strength and can lead to secondary failures like buckling or collapse. Mitigation strategies include resin toughening, interleaving, and incorporating z-pinning or 3D woven fabrics (Saiki and Gomes, 2024).

1.5.2 Fatigue

Fatigue in composites arises from cyclic or fluctuating loads over time. Unlike metals, their anisotropic and heterogeneous structure leads to more complex fatigue behavior. Damage typically initiates as matrix micro-cracks, fiber breakage, or interfacial debonding, eventually progressing to delamination and full structural failure. Fatigue performance depends on several factors:

- Fiber type and orientation
- Matrix toughness
- Stress ratio and amplitude
- Environmental conditions, e.g. (humidity, temperature)
- Manufacturing defects and voids

Given their exposure to prolonged cyclic loading, fatigue testing is critical in aerospace, automotive, and marine sectors (Fazlali et al., 2024).

1.5.3 Matrix Cracking

Matrix cracking is among the earliest damage modes in fiber-reinforced composites, typically arising under transverse tensile loading or from thermal mismatch between the matrix and reinforcement. It involves the initiation and propagation of cracks within the matrix, which may not initially compromise structural integrity but can lead to:

- Stress concentration,
- Fiber-matrix debonding,
- Moisture ingress,
- Progressive degradation of interlaminar bonding.

These cracks often precede delamination or fiber breakage under combined loading. Their initiation and growth depend on matrix toughness, fiber dispersion quality, and the thermal–mechanical load history during service (Jang et al., 2025).

1.6 Applications of Composite Materials

Composite materials are widely used across diverse industries due to their exceptional strength-to-weight ratio, corrosion resistance, and design flexibility. Key applications include:

Aerospace: Composites are extensively used in aircraft structures—including fuselages, engine components, and wings—due to their high strength-to-weight ratio and corrosion resistance. Integrated aerospace designs, such as that of the F-22 fighter, reduce weight and improve manufacturing efficiency. Automated processes like tape-laying (ATL) and fiber placement (AFP) further enhance accuracy and lower assembly costs. ATL, evolved from hand layup, employs robotic or multi-axis systems to place mechanized tapes or dry carbon bundles with consistent compaction (Asyraf et al., 2022; Alam et al., 2022).

Automotive: Composites are employed in car bodies, chassis, and interiors to reduce weight and enhance fuel efficiency. Automated placement technologies, such as Automated Tape Laying (ATL) and Automated Fiber Placement (AFP), enable precise fabrication and integration of material and structure. ATL is ideal for large, low-curvature parts, while AFP suits complex, high-curvature components (Wazeer et al., 2023; Khan et al., 2024).

(a) Construction: Composites are widely applied in bridges, buildings, and pipelines due to their durability and resistance to environmental degradation (Bakis et al., 2002; Mohan et al., 2021).

(b) Renewable Energy: Composites are employed in solar panels and wind turbine blades due to their high strength and lightweight characteristics, enhancing structural efficiency and energy performance (Trzepieciński et al., 2023).

Sports Equipment: Composites are utilized in tennis rackets, golf clubs, and bicycles due to their high stiffness and excellent impact resistance, enhancing performance and durability (Fan et al., 2024).

Environmental Application: Composites such as green composite membranes and mixed metal oxides are employed in air purification due to their effective pollutant removal capabilities (Verma et al., 2022). They are also widely applied in water purification systems for enhanced filtration efficiency and environmental safety (Jaspal and Malviya, 2020).

1.7 Challenges and Future Directions

Despite their advantages, composites face notable challenges:

(a) **High Manufacturing Costs:** Specialized processes and costly raw materials make production more expensive than conventional alternatives.

Recycling Barriers: Multi-material composition hinders recyclability, raising environmental concerns.

(b) **Delamination Risk:** Layer separation under stress compromises structural integrity.

Moisture Sensitivity: Absorption in polymer matrices, driven by temperature and humidity, affects performance. Equilibrium immersion (ASTM D5229) helps simulate worst-case scenarios by ensuring uniform moisture distribution. Emerging materials such as biochar composites and metal-organic frameworks (MOFs), alongside advancements in manufacturing techniques, aim to address existing composite challenges (Ray et al., 2023). Progress in computational modeling further enhances the ability to predict composite behavior under varying conditions (Cheng et al., 2024).

1.8 Conclusion

Composite materials have transformed material science with their exceptional strength, durability, and adaptability, enabling advancements across sectors such as aerospace, automotive, renewable energy, and sports. Despite challenges like high production costs and recycling hurdles, ongoing research is driving the development of more efficient and sustainable alternatives. To align with global sustainability goals, policy recommendations include commercializing biobased composites, establishing recycling protocols, offering sustainability incentives, and fostering industry–academia collaboration. These initiatives support progress toward the Sustainable Development Goals—particularly Goal 9 (Industry, Innovation, and Infrastructure), Goal 12 (Responsible Consumption and Production), and Goal 13 (Climate Action).

References

Alam, M.A., Ya, H.H., Sapuan, S.M., Mamat, O., Parveez, B., Yusuf, M., Masood, F. and Ilyas, R.A. 2022. Recent advancements in advanced composites for aerospace applications: A review. In: *Advanced Composites in Aerospace Engineering Applications*. Springer International Publishing, Cham, pp. 319–339. https://doi.org/10.1007/978-3-030-88192-4_16.

Alam, M.M., Maniruzzaman, M. and Morshed, M.M. 2014. Application and advances in microprocessing of natural fiber (jute)–based composites. *Comprehensive Materials Processing: Thirteen Volume Set*, 7: 243–260. https://doi.org/10.1016/B978-0-08-096532-1.00714-7.

Alves, M., Carlstedt, D., Ohlsson, F., Asp, L.E. and Pimenta, S. 2020. Ultra-strong and stiff randomly-oriented discontinuous composites: Closing the gap to quasi-isotropic continuous-fibre laminates. *Compos. Part A Appl. Sci. Manuf.*, 132: 105826. https://doi.org/10.1016/J.COMPOSITESA.2020.105826.

Anand, A., Ghosh, S.K., Fulmali, A.O. and Prusty, R.K. 2021. Enhanced barrier, mechanical and viscoelastic properties of graphene oxide embedded glass fibre/epoxy composite for marine applications. *Constr. Build Mater.*, 268: 121784. https://doi.org/10.1016/J.CONBUILDMAT.2020.121784.

Asyraf, M.R.M., Ilyas, R.A., Sapuan, S.M., Harussani, M.M., Hariz, H.M., Aiman, J.M., Baitaba, D.M., Sanjay, M.R., Ishak, M.R., Norkhairunnisa, M., Sharma, S., Alam, M.A. and Asrofi, M. 2022. Advanced composite in aerospace applications: opportunities, challenges, and future perspective. In: *Advanced Composites in Aerospace Engineering Applications*. Springer International Publishing, Cham, pp. 471–498. https://doi.org/10.1007/978-3-030-88192-4_24.

Bakis, C.E., Bank, L.C., Brown, V.L., Cosenza, E., Davalos, J.F., Lesko, J.J., Machida, A., Rizkalla, S.H. and Triantafillou, T.C. 2002. Fiber-reinforced polymer composites for construction—state-of-the-art review. *Journal of Composites for Construction*, 6: 73–87. https://doi.org/10.1061/(ASCE)1090-0268(2002)6:2(73).

Bandaru, A.K., Weaver, P.M. and O' Higgins, R.M. 2021. Abrasive wear performance of hygrothermally aged glass/PTFE composites. *Polym. Test*, 103: 107369. https://doi.org/10.1016/J.POLYMERTESTING.2021.107369.

Bhong, M., Khan, T.K.H., Devade, K., Vijay Krishna, B., Sura, S., Eftikhaar, H.K., Pal Thethi, H. and Gupta, N. 2023. Review of composite materials and applications. *Mater. Today Proc.* https://doi.org/10.1016/j.matpr.2023.10.026.

Bokov, D., Turki Jalil, A., Chupradit, S., Suksatan, W., Javed Ansari, M., Shewael, I.H., Valiev, G.H. and Kianfar, E. 2021. Nanomaterial by sol-gel method: synthesis and application. *Advances in Materials Science and Engineering*, 2021. https://doi.org/10.1155/2021/5102014.

Carey, J.P., Melenka, G.W., Hunt, A.J., Lepp, E., Pastore, C.M. and Ko, F.K. 2024. Advanced testing of braided composite materials. In: *Handbook of Advances in Braided Composite Materials*. Elsevier, pp. 161–210. https://doi.org/10.1016/B978-0-443-18602-8.00016-4.

Cheang, P. and Khor, K.A. 1996. Addressing processing problems associated with plasma spraying of hydroxyapatite coatings. *Biomaterials*, 17: 537–544. https://doi.org/10.1016/0142-9612(96)82729-3.

Cheng, Z.-Q., Liu, H. and Tan, W. 2024. Advanced computational modelling of composite materials. *Eng. Fract. Mech.*, 305: 110120. https://doi.org/10.1016/j.engfracmech.2024.110120.

Choudhury, M., Singh Bindra, H., Mittal, J. and Nayak, R. 2021. Evaluation of mechanical properties of carbon HDPE composites. *Mater. Today Proc.*, 47: 6712–6718. https://doi.org/10.1016/J.MATPR.2021.05.119.

de Castro Saiki, L.E. and Gomes, G.F. 2024. Understanding and mitigating delamination in composite materials: A comprehensive review. *Mechanics of Advanced Materials and Structures*, 31: 13147–13167. https://doi.org/10.1080/15376494.2024.2333490.

Demircioglu, T.K., Balikoglu, F., Beyaz, S. and Bülbül, B. 2021. Effect of lead metaborate as novel nanofiller on the ballistic impact behavior of Twaron®/epoxy composites. *Composites Communications*, 27: 100832. https://doi.org/10.1016/J.COCO.2021.100832.

Devasia, R., Painuly, A., Devapal, D. and Sreejith, K.J. 2021. Continuous fiber reinforced ceramic matrix composites. In: *Fiber Reinforced Composites*. Elsevier, pp. 669–751. https://doi.org/10.1016/B978-0-12-821090-1.00022-3.

Diniță, A., Ripeanu, R.G., Ilincă, C.N., Cursaru, D., Matei, D., Naim, R.I., Tănase, M. and Portoacă, A.I. 2023. Advancements in fiber-reinforced polymer composites: A comprehensive analysis. *Polymers (Basel)*, 16: 2. https://doi.org/10.3390/polym16010002.

Etri, H. EL, Polat, M.Z. and Özyılmaz, E.B. 2024. Classification, applications, and impact behavior of composite materials: A review. *Mechanics of Advanced Materials and Structures*, 31: 12592–12635. https://doi.org/10.1080/15376494.2024.2326665.

Fan, Q., Duan, H. and Xing, X. 2024. A review of composite materials for enhancing support, flexibility and strength in exercise. *Alexandria Engineering Journal*, 94: 90–103. https://doi.org/10.1016/j.aej.2024.03.048.

Fazlali, B., Lomov, S.V. and Swolfs, Y. 2024. Reducing stress concentrations in static and fatigue tensile tests on unidirectional composite materials: A review. *Compos. B Eng.*, 273: 111215. https://doi.org/10.1016/j.compositesb.2024.111215.

Fernandes, C., Pontes, A.J., Viana, J.C. and Gaspar-Cunha, A. 2018. Modeling and optimization of the injection-molding process: A review. *Advances in Polymer Technology*, 37: 429–449. https://doi.org/10.1002/adv.21683.

Fong, L. and Advani, S.G. 1998. Resin transfer molding. In: *Handbook of Composites*. Springer US, Boston, MA, pp. 433–455. https://doi.org/10.1007/978-1-4615-6389-1_21.

Glossary of Filament-Winding Terms. 2011. In: *Composite Filament Winding*. ASM International, pp. 151–158. https://doi.org/10.31399/asm.tb.cfw.t52860151.

Haluza, R.T., Koudela, K.L., Bakis, C.E., Adams, D.O. and Pereira, J.M. 2022. Shear strain measurement techniques in composite v-notch shear testing. *Exp. Mech.*, 62: 1655–1671. https://doi.org/10.1007/s11340-022-00897-9.

Hasan, M.M.B., Abdkader, A., Cherif, C. and Spennato, F. 2019a. Fibre hybrid composites consisting of discontinuous waste carbon fibre and continuous glass filaments developed for load-bearing structures with improved impact strength. *Compos. Part A Appl. Sci. Manuf.*, 126: 105610. https://doi.org/10.1016/J.COMPOSITESA.2019.105610.

Hasan, M.M.B., Abdkader, A., Cherif, C. and Spennato, F. 2019b. Fibre hybrid composites consisting of discontinuous waste carbon fibre and continuous glass filaments developed for load-bearing structures with improved impact strength. *Compos. Part A Appl. Sci. Manuf.*, 126: 105610. https://doi.org/10.1016/J.COMPOSITESA.2019.105610.

Jaafar, J., Siregar, J.P., Tezara, C., Hamdan, M.H.M. and Rihayat, T. 2019. A review of important considerations in the compression molding process of short natural fiber composites. *The International Journal of Advanced Manufacturing Technology*, 105: 3437–3450. https://doi.org/10.1007/s00170-019-04466-8.

Jang, D., Park, J., Choi, S., Bang, J., Choi, J., Kim, J., Yang, B. and Jeon, H. 2025. Novel approach for crack detections and rapid repairment methods in cement-based self-heating composites for smart infrastructures. *Compos. B Eng.*, 293: 112126. https://doi.org/10.1016/j.compositesb.2025.112126.

Jaspal, D. and Malviya, A. 2020. Composites for wastewater purification: A review. *Chemosphere*, 246: 125788. https://doi.org/10.1016/j.chemosphere.2019.125788.

Kargarzadeh, H., Mariano, M., Huang, J., Lin, N., Ahmad, I., Dufresne, A. and Thomas, S. 2017a. Recent developments on nanocellulose reinforced polymer nanocomposites: A review. *Polymer (Guildf)*. https://doi.org/10.1016/j.polymer.2017.09.043.

Kargarzadeh, H., Mariano, M., Huang, J., Lin, N., Ahmad, I., Dufresne, A. and Thomas, S. 2017b. Recent developments on nanocellulose reinforced polymer nanocomposites: A review. *Polymer (Guildf)*, 132: 368–393. https://doi.org/10.1016/j.polymer.2017.09.043.

Khan, F., Hossain, N., Mim, J.J., Rahman, S.M., Iqbal, Md.J., Billah, M. and Chowdhury, M.A. 2024. Advances of composite materials in automobile applications—A review. *Journal of Engineering Research*. https://doi.org/10.1016/j.jer.2024.02.017.

Ku, H., Wang, H., Pattarachaiyakoop, N. and Trada, M. 2011. A review on the tensile properties of natural fiber reinforced polymer composites. *Compos. B Eng.*, 42: 856–873. https://doi.org/10.1016/j.compositesb.2011.01.010.

Lee, S.-Y. and Springer, G.S. 1990. Filament winding cylinders. *J. Compos. Mater.*, 24: 1270–1298. https://doi.org/10.1177/002199839002401202.

Li, R., Xia, L., Lyu, P., Zhang, J., Wang, Y., Deng, B., Zhang, C., Liu, X. and Xu, W. 2021. High content filling, toughness, and conductive performance of thermoplastic polyurethane/carbon nanotubes composites prepared by constructing the compact interface. *Composites Communications*, 28: 100948. https://doi.org/10.1016/J.COCO.2021.100948.

Liang, H., Xu, M., Chen, X., Bu, Y., Zhang, Y., Gao, K., Min, C., Hua, X. and Fu, Y. 2021. Impressive high-temperature lubrication of carbon/epoxy composites endowed by the microscopic dimensionality of carbon fillers. *Wear*, 486-487: 204109. https://doi.org/10.1016/J.WEAR.2021.204109.

Liu, M., Chen, B., Zhang, D., Wang, Y. and Kan, P. 2021. Numerical studies on effective thermal conductivities of the glass/polyimide composite materials under the conditions of conduction & radiation. *Int. J. Heat Mass Transf.*, 180: 121764. https://doi.org/10.1016/J.IJHEATMASSTRANSFER.2021.121764.

Lo, J. and Nutt, S. 2019. Method for *in situ* analysis of volatiles generated during cure of composites. *Compos. Part A Appl. Sci. Manuf.*, 123: 141–148. https://doi.org/10.1016/j.compositesa.2019.05.013.

Middleton, B. 2016. Composites: Manufacture and application. In: *Design and Manufacture of Plastic Components for Multifunctionality*. Elsevier, pp. 53–101. https://doi.org/10.1016/B978-0-323-34061-8.00003-X.

Mohan, H.T., Jayanarayanan, K. and Mini, K.M. 2021. Recent trends in utilization of plastics waste composites as construction materials. *Constr. Build. Mater.*, 271: 121520. https://doi.org/10.1016/j.conbuildmat.2020.121520.

Morwenna Spear. 2018. Natural Materials—Composition and Combinations. In: *Designing with Natural Materials*. CRC Press, p. 82.

Moschiar, S.M., Reboredo, M.M. and Vazquez, A. 1998. Pultrusion processing. In: *Advanced Polymer Processing Operations*. Elsevier, pp. 126–156. https://doi.org/10.1016/B978-081551426-8.50009-X.

Muthalagu, R., Murugesan, J., Sathees Kumar, S. and Sridhar Babu, B. 2021. Tensile attributes and material analysis of kevlar and date palm fibers reinforced epoxy composites for automotive bumper applications. *Mater. Today Proc.*, 46: 433–438. https://doi.org/10.1016/J.MATPR.2020.09.777.

Neelesh K. Jain, Mayur S. Sawant, Sagar H. Nikam and Suyog Jhavar. 2016. Metal deposition: Plasma-based processes. In: *Encyclopedia of Plasma Technology*. CRC Press, pp. 722–740. https://doi.org/10.1081/E-EPLT-120053919.

Neher, B., Bhuiyan, Md.M.R., Kabir, H., Qadir, Md.R., Gafur, Md.A. and Ahmed, F. 2014. Study of mechanical and physical properties of palm fiber reinforced acrylonitrile butadiene styrene composite. *Materials Sciences and Applications*, 05: 39–45. https://doi.org/10.4236/msa.2014.51006.

Nicharat, A., Sapkota, J., Weder, C. and Foster, E.J. 2015. Melt processing of polyamide 12 and cellulose nanocrystals nanocomposites. *J. Appl. Polym. Sci.*, 132. https://doi.org/10.1002/app.42752.

Pan, L., Guo, H., Zhong, L., Wang, M., Xue, P. and Yuan, X. 2021. Influence of surface-modified glass fibers on interfacial properties of GF/PEEK composites using molecular dynamics. *Comput. Mater. Sci.*, 188: 110216. https://doi.org/10.1016/J.COMMATSCI.2020.110216.

Pascault, J.P. and Williams, R.J.J. 2018. Overview of thermosets: Present and future. *Thermosets: Structure, Properties, and Applications*: Second Edition, 3–34. https://doi.org/10.1016/B978-0-08-101021-1.00001-0.

Patel, P. and Chaudhary, V. 2022. Delamination evaluation in drilling of composite materials—A review. *Mater. Today Proc.*, 56: 2690–2695. https://doi.org/10.1016/J.MATPR.2021.09.267.

Pecić, N., Mašović, S. and Stošić, S. 2023. Span-to-depth ratio limits for deflection control of reinforced concrete elements. *Structural Concrete,* 24: 2547–2562. https://doi.org/10.1002/suco.202200132.

Pilato, L.A. and Michael J. Michno. 1994. *Advanced Composite Materials*. Springer Science & Business Media.

Pilato, L.A. and Michno, M.J. 1994. *Advanced Composite Materials*. Springer Berlin Heidelberg.

Plastics Europre. 2025. Thermoplastics • Plastics Europe [WWW Document]. https://plasticseurope.org/plastics-explained/a-large-family/thermoplastics/.URL https://plasticseurope.org/plastics-explained/a-large-family/thermoplastics/ (accessed 4.3.25).

Ramesh, M., Tamil Selvan, M. and Saravanakumar, A. 2025. Evolution and recent advancements of composite materials in structural applications. In: *Applications of Composite Materials in Engineering*. Elsevier, pp. 97–117. https://doi.org/10.1016/B978-0-443-13989-5.00004-8.

Ray, M., Verma, A., Maiti, A. and Negi, Y.S. 2023. Nano-engineered polymer matrix-based composites. In: *Polymer Nanocomposites*. CRC Press, Boca Raton, pp. 21–39. https://doi.org/10.1201/9781003343912-2.

Ren, J., Chen, L., Liu, Z., Song, Q. and Liu, C. 2021. Study on the heat transfer reinforcement of glass fiber/epoxy resin composites by grafting and dispersing graphene oxide. *Compos. Sci. Technol.*, 216: 109039. https://doi.org/10.1016/J.COMPSCITECH.2021.109039.

Repka, M.A., Shah, S., Lu, J., Maddineni, S., Morott, J., Patwardhan, K. and Mohammed, N.N. 2012. Melt extrusion: Process to product. *Expert. Opin. Drug Deliv.*, 9: 105–125. https://doi.org/10.1517/17425247.2012.642365.

Sakka, S. 2016. History of the sol–gel chemistry and technology. In: *Handbook of Sol-Gel Science and Technology*. Springer International Publishing, Cham, pp. 1–27. https://doi.org/10.1007/978-3-319-19454-7_87-1.

Sano, Y., Matsuzaki, R., Ueda, M., Todoroki, A. and Hirano, Y. 2018. 3D printing of discontinuous and continuous fibre composites using stereolithography. *Addit. Manuf.*, 24: 521–527. https://doi.org/10.1016/J.ADDMA.2018.10.033.

Shrikaant Kulkarni. 2019. Characterization techniques for polymers and polymer nanocomposites. In: *Composite Materials Engineering*. Apple Academic Press, p. 50.

Sim, J., Yu, S.-H., Yoon, H.-S., Kwon, D.-J., Lee, D.-H. and Bae, J.-S. 2021. Characteristic evaluation and finite element analysis of glass fiber/recycled polyester thermoplastic composites by cross-sectional shape of glass fiber. *Compos. B Eng.*, 223: 109095. https://doi.org/10.1016/J.COMPOSITESB.2021.109095.

Sivasubramanian, P., Mayandi, K., Arumugaprabu, V., Rajini, N. and Rajesh, S. 2021. History of composites and polymers. In: *Polymer-Based Composites*. CRC Press, pp. 1–21.

Smith, A.W. and Hubert, P. 2023. Benchmarking of potential rapid inspection methods for quantitative tracking of thermoset prepreg state evolution with prolonged ambient ageing. *Compos. B Eng.*, 254: 110579. https://doi.org/10.1016/j.compositesb.2023.110579.

Stark, W. 2013. Investigation of the curing behaviour of carbon fibre epoxy prepreg by Dynamic Mechanical Analysis DMA. *Polym, Test*, 32: 231–239. https://doi.org/10.1016/j.polymertesting.2012.11.004.

Trzepieciński, T., Batu, T., Kibrete, F. and Lemu, H.G. 2023. Application of composite materials for energy generation devices. *Journal of Composites Science*, 7: 55. https://doi.org/10.3390/jcs7020055.

Verma, A., Mishra, N.C. and Maiti, A. 2022. Green composite membranes in air purification. 1st Editio. ed, *Encyclopedia of Green Materials*. Springer Nature Singapore, Singapore. https://doi.org/10.1007/978-981-16-4921-9_259-1.

Vieille, B., Coppalle, A., Schuhler, E., Chaudhary, A., Alia, A., Delpouve, N. and Bourdet, A. 2022. Influence of kerosene flame on fire-behaviour and mechanical properties of hybrid Carbon

Glass fibers reinforced PEEK composite laminates. *Compos. Struct.*, 279: 114786. https://doi.org/10.1016/J.COMPSTRUCT.2021.114786.

Vita, A., Castorani, V., Germani, M. and Marconi, M. 2019. Comparative life cycle assessment and cost analysis of autoclave and pressure bag molding for producing CFRP components. *The International Journal of Advanced Manufacturing Technology*, 105: 1967–1982. https://doi.org/10.1007/s00170-019-04384-9.

Wazeer, A., Das, A., Abeykoon, C., Sinha, A. and Karmakar, A. 2023. Composites for electric vehicles and automotive sector: A review. *Green Energy and Intelligent Transportation*, 2: 100043. https://doi.org/10.1016/j.geits.2022.100043.

Wu, C., Xu, F., Wang, H., Liu, H., Yan, F. and Ma, C. 2023. Manufacturing technologies of polymer composites—A review. *Polymers*, 15: 712. https://doi.org/10.3390/POLYM15030712.

Xu, T., Zhou, S., Jiang, F., Song, N., Shi, L. and Ding, P. 2021. Polyamide composites with improved thermal conductivity for effective thermal management: The three-dimensional vertically aligned carbon network. *Compos. B Eng.*, 224: 109205. https://doi.org/10.1016/J.COMPOSITESB.2021.109205.

Xue, L., Fan, W., Wu, F., Zhang, Y., Guo, K., Li, J., Yuan, L., Dang, W. and Sun, R. 2019. The influence of thermo-oxidative aging on the electromagnetic absorbing properties of 3D quasi-isotropic braided carbon/glass bismaleimide composite. *Polym. Degrad. Stab.*, 168: 108941. https://doi.org/10.1016/J.POLYMDEGRADSTAB.2019.108941.

Yamaguchi, N., Tadanaga, K., Matsuda, A., Minami, T. and Tatsumisago, M. 2006. Formation of anti-reflective alumina films on polymer substrates by the sol–gel process with hot water treatment. *Surf. Coat. Technol.*, 201: 3653–3657. https://doi.org/10.1016/j.surfcoat.2006.08.122.

Zhang, X., Ma, J., Yan, R., Cheng, W., Zheng, J. and Jin, B. 2021. Pt-Ru/polyaniline/carbon nanotube composites with three-layer tubular structure for efficient methanol oxidation. *J. Alloys Compd.*, 867: 159017. https://doi.org/10.1016/J.JALLCOM.2021.159017.

Zhao, L., Mantell, S.C., Cohen, D. and McPeak, R. 2001. Finite element modeling of the filament winding process. *Compos. Struct.*, 52: 499–510. https://doi.org/10.1016/S0263-8223(01)00039-3.

Zhu, L., Zhu, G., Feng, J., Jin, L. and Ma, P. 2019. Impact tensile behavior analysis of Twaron fiber tows from fast Fourier transform. *Textile Research Journal*, 89: 1363–1370. https://doi.org/10.1177/0040517518775917/ASSET/89AAB72C-5538-40D4-964D-CD6E2670A83A/ASSETS/IMAGES/LARGE/10.1177_0040517518775917-FIG9.JPG.

CHAPTER

2 Emerging Trends in Advanced Hybrid Materials

Innovations and Future Prospects

Sanjeev Kumar Sharma,[1,*] *Sudhir Agrawal*,[2]
Darshan Srivastav,[1] *Vinod Baniya*[1]
and *Disha Mishra*[1]

Advanced hybrid materials are driving innovation across sectors like agriculture, healthcare, energy, and electronics by combining diverse material types to create synergistic properties. Their multifunctionality makes them ideal for applications in automotive, structural, and biomedical fields. Fueled by progress in nanoscience and nanotechnology, hybrid materials—unlike traditional composites—often incorporate nanoscale components, broadening their functional scope and impact across modern industries.

This chapter presents an overview of the hybrid materials landscape, linking key subfields, synthesis methods, and applications. It emphasizes hybridization as a catalyst for technological progress and explores its role in sustainable solutions—such as green construction, renewable energy, and eco-friendly packaging—to support both economic and environmental goals.

[1] Department of Applied Science & Humanities, Maharana Pratap Institute of Technology, Gorakhpur, India.

[2] Department of Computer Science and Engineering, Maharana Pratap Institute of Technology, Gorakhpur, India.

* Corresponding author: jms.sanjeev@gmail.com

ORCID IDs of authors: https://orcid.org/0000-0001-9694-146X; https://orcid.org/0009-0002-7658-3508; https://orcid.org/0000-0002-0585-5767; https://orcid.org/0000-0002-5667-9265; https://orcid.org/0009-0009-8116-5678

DOI: 10.1201/9781003607359-2

2.1 Introduction

Hybrid materials have gained significant interest in recent decades for combining organic and inorganic components at the molecular or nanoscale. This integration yields synergistic properties—such as improved strength, thermal stability, optical activity, conductivity, and chemical resilience—that surpass those of the individual components.

Hybrid materials are broadly classified into two types based on organic–inorganic interactions (Porcu et al., 2022; Jyoti et al., 2022; Gomez-Romero et al., 2023; Gao et al., 2023; Ali et al., 2024). Class I hybrids rely on weak forces—hydrogen bonding, van der Waals, or electrostatic—preserving individual component traits while enhancing dispersion and interface compatibility. Class II hybrids feature strong covalent or ionic–covalent bonds, creating new structures and often unique functionalities (Liu et al., 2021; Zhou et al., 2021; Zhu et al., 2021; Sun et al., 2022; Xing et al., 2022).

Powered by advances in nanotechnology and materials chemistry, hybrid materials are now widely used in energy storage, catalysis, environmental cleanup, and biomedicine. In lithium-ion batteries and supercapacitors, they boost electrochemical performance through improved conductivity and structural strength. Their photocatalytic properties also support efficient pollutant breakdown and energy conversion in environmental systems. In biomedicine, hybrid nanoparticles are designed for targeted drug delivery, biosensing, and implantable coatings. They enable controlled drug release with enhanced biocompatibility and lower toxicity. Additionally, they show strong potential in high-sensitivity detection for diagnostics and environmental monitoring. Despite their promise, hybrid materials face key challenges—limited scalability, high production costs, poor long-term stability, and recyclability issues—hindering widespread industrial and environmental adoption. This chapter provides a structured overview of hybrid materials, covering their classification, synthesis, applications, and emerging trends. It highlights their pivotal role in advancing sustainable technologies, linking scientific innovation to real-world solutions.

2.2 Classification of Hybrid Materials

Hybrid materials are classified by their component makeup and the interactions that bind them, which is key to understanding their properties, performance, and application potential. The most widely accepted model, introduced by Judeinstein and Sanchez, divides hybrid materials into two main classes:

- **Class I Hybrids**

  Class I hybrids rely on weak interactions—hydrogen bonding, van der Waals forces, or electrostatic attraction—allowing each component to retain its identity. While offering better dispersion and compatibility, they are easier to synthesize but less structurally stable under harsh conditions.

- **Class II Hybrids**

  Class II hybrids feature strong covalent or ionic–covalent bonds between organic and inorganic components, forming new structures with distinct mechanical, thermal, or optical properties not present in the individual materials. While the binary classification offers clarity, it's not absolute. Many advanced hybrids exhibit both weak and strong interactions, resulting in complex architectures. A more refined system categorizes them by matrix and filler roles into four types:

- I–O (Inorganic matrix–Organic filler)
- O–I (Organic matrix–Inorganic filler)
- I–I (Inorganic–Inorganic hybrid systems)
- O–O (Organic–Organic hybrids with distinct functionalities)

This multi-level classification offers deeper insight into hybrid material structures and guides application-specific design in industrial, environmental, and biomedical fields. Figure 2.1 illustrates this dual scheme: interaction-based (Class I and II) on the left, and component-based (I–O, O–I, I–I, O–O) on the right.

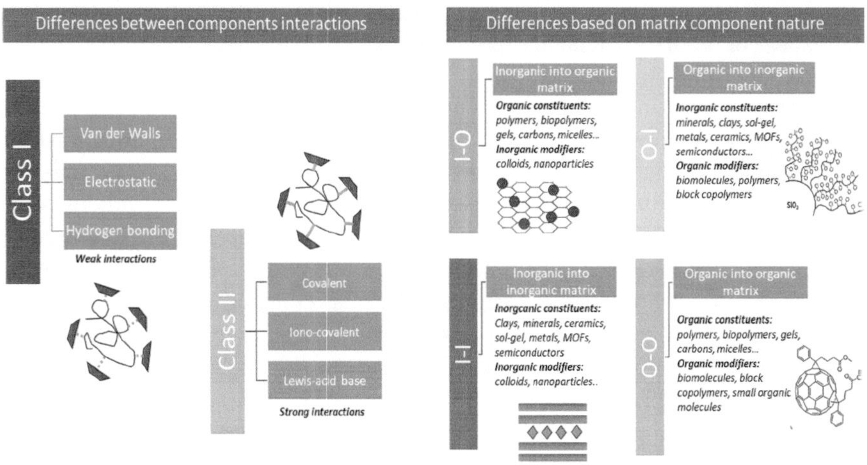

Figure 2.1: General classification of hybrid materials. Reproduced with permission from (Gomez-Romero et al., 2023).

2.2.1 Organic-Inorganic Hybrid Materials

Organic–inorganic hybrid materials blend the robustness of metals, oxides, or ceramics with the flexibility and tunability of organics like polymers or small molecules. This synergy creates advanced materials with superior properties beyond those of their individual components. Hybridization usually takes place at the molecular or nanoscale level, enabling close contact and strong synergy between organic and inorganic components. This enhances mechanical, thermal, chemical, electrical, and optical properties. By adjusting composition, structure, and bonding, hybrids can be tailored for specific applications.

The sol–gel process is a widely used method for embedding organic molecules into inorganic matrices like silica or metal oxides. It produces hybrids with controlled porosity, high optical clarity, and strong mechanical properties—suitable for thin films, optics, and filtration. Another key category is *metal-organic frameworks* (MOFs)—crystalline structures of metal ions and organic ligands with high porosity and modularity. Their tunable pore size, surface properties, and structure make them ideal for gas adsorption, separation, catalysis, and drug delivery. *Organosilica materials*, with silicon–oxygen backbones and organic side groups, offer better thermal and hydrolytic stability than traditional silicates. They show promise in catalysis, chromatographic separation, and nanomedicine. Thanks to their multifunctionality and design versatility, organic–inorganic hybrids play a key role in advancing electronics, energy, photonics, green construction, and biomedicine. Their unique blend of robustness and adaptability places them at the forefront of modern materials innovation.

2.2.2 Composite Hybrid Materials

Composite hybrid materials combine two or more distinct phases—organic, inorganic, or both—to enhance performance. Unlike molecular-level bonded hybrids, these may involve physical mixing or chemical integration. Their design focuses on maximizing each component's strengths while minimizing drawbacks. The tailored composition and microstructure of composite hybrids enable outstanding strength, thermal resistance, conductivity, and durability—making them ideal for high-performance uses in electronics, aerospace, automotive, environmental cleanup, and energy storage. A major advancement in this field is the incorporation of carbon nanomaterials like CNTs and graphene into polymers, significantly boosting tensile strength, thermal conductivity, and electrical performance. CNT–polymer composites, in particular, are valued for their lightweight, flexibility, and conductivity—ideal for flexible electronics, sensors, and coatings. Graphene oxide–metal oxide composites combine graphene's high surface area and conductivity with the functional traits of metal oxides (e.g., TiO_2, ZnO, Fe_2O_3). Their efficient charge transfer and structural stability make them ideal for photocatalysis, gas sensing, and energy conversion.

Polymer–carbon composites pair conductive carbon fillers with insulating or semiconducting polymers to create materials suited for EMI shielding, thermal control, and battery electrodes. Polymers offer easy processing and moldability, while carbon boosts electrical and thermal performance. Fine-tuning microstructure and interfaces in composite hybrids paves the way for next-gen materials tailored to specific performance needs. As demand rises for multifunctional, lightweight, and robust systems, these hybrids remain central to innovation in both research and industry. Figure 2.2 shows graphene-based materials for advanced energy storage and related applications.

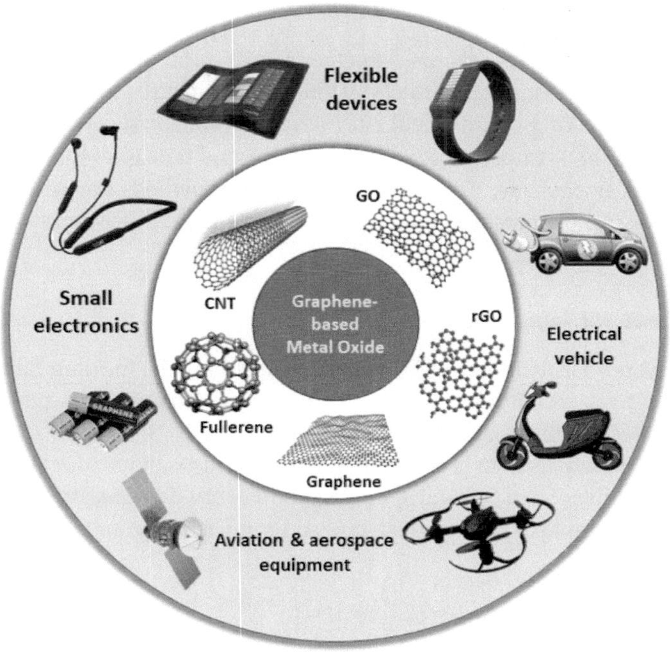

Figure 2.2: Overview of graphene-based materials for advanced energy storage devices and their potential applications in various fields. Reproduced with permission from (Potbhare et al., 2024).

2.3 Synthesis of Hybrid Materials

Synthesis methods are key to shaping the structure, properties, and performance of hybrid materials. Fabrication techniques influence critical factors like particle size, morphology, porosity, crystallinity, and phase dispersion, all of which determine application suitability.

An essential goal in hybrid material synthesis is achieving uniform component dispersion, strong interfacial bonding, and tunable structure. Depending on the application—catalysis, energy, environment, or biomedicine—various established synthesis methods are employed.

2.3.1 Sol–Gel Process

The sol–gel method is a widely used technique for hybrid synthesis, involving hydrolysis and condensation of metal precursors in solution to form a gel. After drying and heat treatment, the gel solidifies into a porous or dense hybrid material. A key benefit of the sol–gel method is its mild processing conditions, ideal for including heat-sensitive organic molecules. It yields hybrids with high surface area, adjustable porosity, and uniform composition—widely used in coatings, sensors, catalysts, membranes, and biomedical devices.

2.3.2 Hydrothermal and Solvo-thermal Methods

Hydrothermal and *solvo-thermal* methods use sealed autoclaves at high temperature and pressure to form crystalline nanostructures and intricate hybrid architectures, making them ideal for advanced material synthesis. Tuning parameters like pH, temperature, solvent polarity, and time allows precise control over material size, shape, and composition. These methods are widely used to create MOFs, nanocomposites, and porous oxides for catalysis, gas storage, and environmental cleanup.

2.3.3 Chemical Vapor Deposition (CVD)

Chemical Vapor Deposition (CVD) is a precise method for forming thin films by decomposing gas precursors on heated surfaces, yielding uniform, pure coatings with controlled thickness and strong adhesion.

CVD is widely used to produce graphene hybrids, semiconductor films, and protective coatings for electronics, optics, and energy devices. Its precision and ability to coat complex surfaces make it ideal for microelectronics, solar cells, and aerospace uses.

2.3.4 Electrospinning

Electrospinning produces hybrid nanofibers by applying high voltage to a polymer solution or melt, often with inorganic nanoparticles. This forms fine fibers—nanometers to micrometers in diameter—collected as nonwoven mats.

Electrospun fibers provide high surface area, porosity, and flexibility, making them ideal for tissue engineering, drug delivery, filtration, and energy storage. Adjusting voltage, flow rate, and collector design allows precise control over their morphology and function. Together, these synthesis methods offer researchers flexible tools to engineer hybrid materials for targeted performance. With rising demand for versatile, affordable, and scalable solutions, continued innovation in synthesis remains central to materials science progress.

2.4 Innovations in Advanced Hybrid Materials

Ongoing advances in hybrid materials are driving breakthroughs across science and industry. Current research emphasizes smart, adaptable, and eco-friendly systems, with key trends including stimuli-responsive materials, multifunctional nanocomposites, and AI-driven design.

2.4.1 Hybrid Materials for Energy Storage and Conversion

Hybrid materials are driving advances in energy storage and conversion by combining electrochemical, thermal, and mechanical properties. This makes them ideal for next-gen batteries, super capacitors, and solar cells.

In lithium-ion batteries (LIBs), hybrid electrodes combining graphene with metal oxides like MnO_2 or Fe_3O_4 boost capacity, rate performance, and stability. Graphene's conductivity and strength improve charge flow and absorb volume changes during cycling, enhancing overall battery lifespan. Hybrid supercapacitors blend carbon materials with pseudo-capacitive elements like metal oxides or conductive polymers, combining high energy density with long cycle life—ideal for EVs, renewable energy storage, and fast-charging devices. Hybrid materials are advancing solar tech, particularly in perovskite solar cells, where blending polymers with inorganic perovskites enhances efficiency, flexibility, and reduces cost. Figure 2.3 highlights MOF thin films in perovskite and dye-sensitized cells.

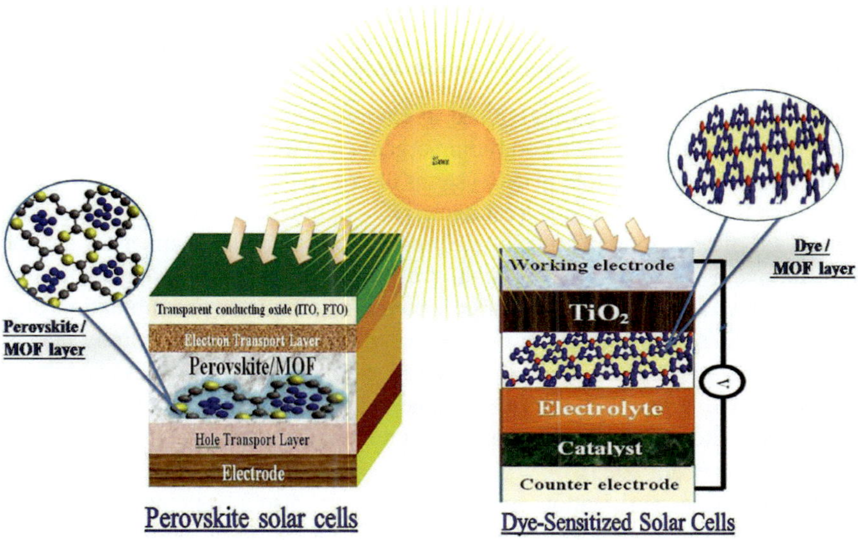

Figure 2.3: The role of MOF thin films in perovskite and dye-sensitized solar cells. Reproduced with permission from (Chuhadiya et al., 2021).

2.4.2 Hybrid Materials for Catalysis

Hybrid materials have transformed catalysis by enabling precise control over surface properties, porosity, and reactivity.

Metal-organic frameworks (MOFs) are standout catalysts due to their high surface area and tunable pores, making them ideal for CO_2 capture, biomass conversion, and fine chemical synthesis.

Hybrid photo-catalysts—made by combining semiconductors like TiO_2 with dyes or conductive polymers—boost light absorption and charge separation under visible light. They're used in solar hydrogen production and pollutant degradation, supporting clean energy and environmental solutions. Figure 2.4 shows metal–carbon hybrids and their properties.

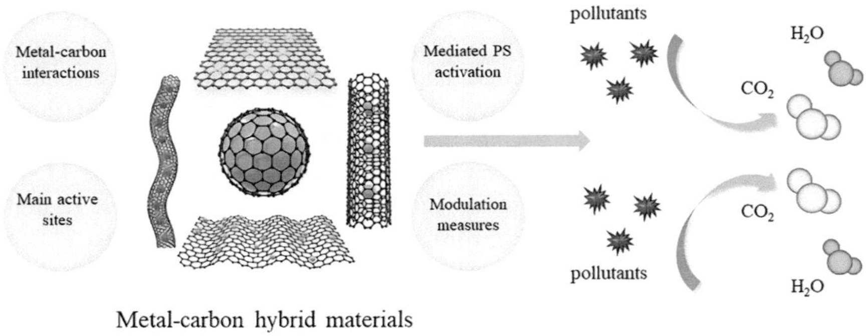

Figure 2.4: Metal-carbon hybrid materials exhibiting promising catalysts properties. Reproduced with permission from (Yan Huchuan, 2023).

2.4.3 Hybrid Materials for Environmental Remediation

Hybrid materials are increasingly used in environmental cleanup, especially for water and air purification. By combining organic and inorganic components, they outperform traditional single-phase systems thanks to their multifunctionality and molecular tunability. In water treatment, TiO_2–graphene composites stand out: while TiO_2 offers strong photocatalytic power, adding graphene enhances charge separation and visible light absorption, greatly boosting pollutant degradation in wastewater. Graphene enhances charge separation, minimizes recombination, and broadens TiO_2's light absorption into the visible range. This significantly boosts the photocatalytic degradation of dyes, pharmaceuticals, and pesticides under both UV and visible light—making TiO_2–graphene hybrids ideal for energy-efficient, sustainable water treatment, particularly in underserved regions (Prasannan et al., 2020; Grisolia et al., 2023).

Beyond water purification, hybrid materials are also advancing air filtration by targeting volatile organic compounds (VOCs)—toxic emissions from products like paints and solvents linked to respiratory and long-term health risks. Combining carbon-based substrates (e.g., activated carbon, CNTs) with metal oxides (ZnO, TiO_2, MnO_2), these hybrids offer high surface area and active sites for photocatalytic or thermal VOC degradation into harmless byproducts. They're widely applied in filters, coatings, and purifiers across sectors (Ali et al., 2024; Zhu et al., 2024; Nawaz et al., 2024). Figure 2.5 shows the PAN–carrageenan/laponite hybrid and its multifunctional roles in pollutant adsorption.

The adaptability of hybrid materials enables functionalization or doping to target specific pollutants. For example, nitrogen-doped TiO_2 or TiO_2 coupled with ZnO improves visible-light photocatalysis, vital for indoor settings with limited sunlight. Porous carbon frameworks further boost adsorption and diffusion, enhancing purification efficiency. Relevant materials for VOC applications are shown in Figure 2.6 (Nawaz et al., 2024).

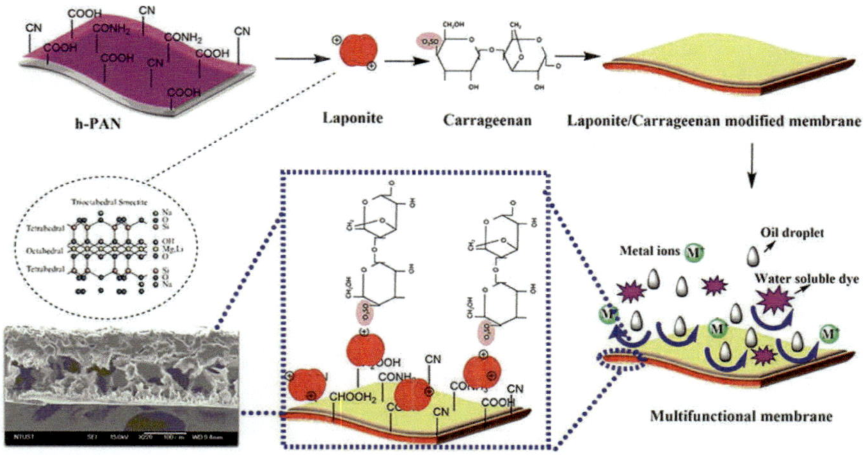

Figure 2.5: Schematic diagram of fabrication method of PAN modified by carrageenan/laponite multilayer, and its multifunctional applications of adsorbents of water pollutants. Reproduced with permission from (Prasannan et al., 2020).

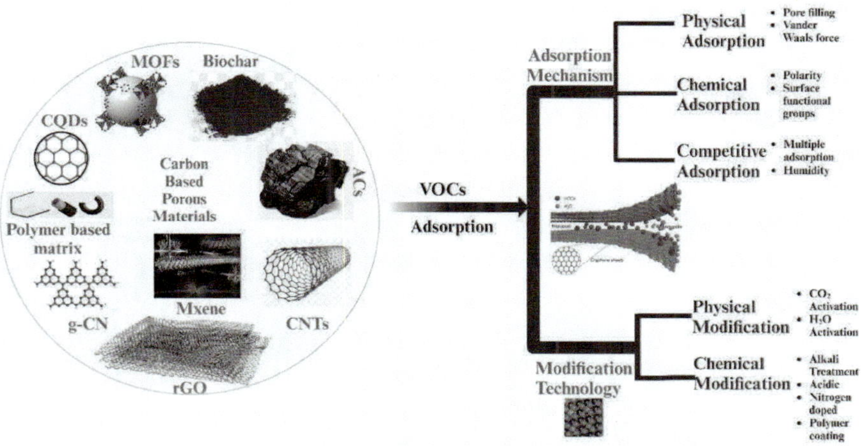

Figure 2.6: Schematic showing the various materials for VOCs applications. Reproduced with permission from (Nawaz et al., 2024).

2.4.4 Hybrid Materials for Sensors and Actuators

Hybrid materials are key to next-gen sensors and actuators by integrating diverse material properties into one functional system. Metal oxide–polymer hybrids—like ZnO, SnO_2, or TiO_2 combined with conductive polymers—are especially effective for gas sensing, offering improved sensitivity, selectivity, and faster response than single-phase sensors. While metal oxides excel in detecting gases via resistance changes, they often require high temperatures and show cross-sensitivity—challenges hybrid systems help overcome (Xu et al., 2024; Sun et al., 2024). To address these challenges, researchers have incorporated conductive polymers like PANI, PPy, and polythiophene into hybrid sensors. These enhance conductivity and introduce more gas interaction sites at lower temperatures. The resulting metal oxide–polymer hybrids exhibit superior sensitivity and selectivity for gases such as NH_3, CO_2, and NO_x, making them vital for safety, diagnostics, and smart systems. Figure 2.7 presents the representative IL compositions, properties, and applications.

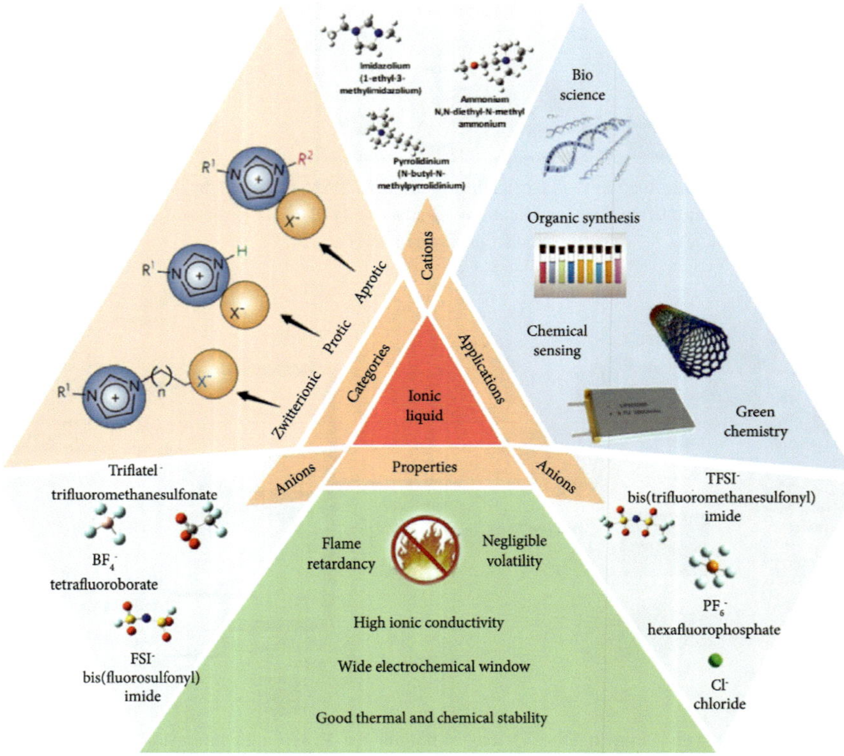

Figure 2.7: Representative compositions, properties, categories, and applications of ILs. Reproduced with permission from (Zhu et al., 2021).

Hybrid materials are being actively developed for actuators, especially in soft robotics where responsiveness to stimuli like temperature, humidity, pH, or light is key. Hybrids combining thermo-responsive polymers with inorganic nanoparticles or photo-responsive composites can bend, expand, or move under specific triggers (Ivanova et al,2024, Fernandes et al,2025). Figure 2.8 illustrates the broad applications of ILs across generations.

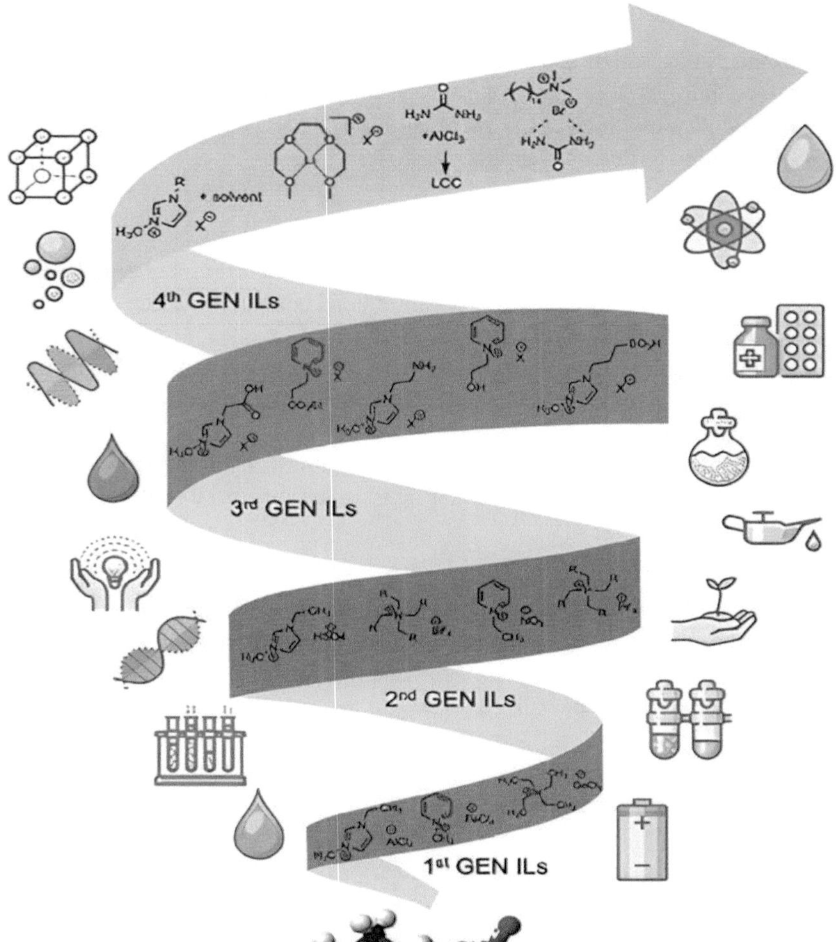

Figure 2.8: Illustrates the evolution of ILs from first to fourth generations, including ionic liquid analogs. Colored ribbons represent each generation with examples, while surrounding icons depict their broad applications—ranging from batteries and supercapacitors to solvents, catalysts, biomedical uses, gas separation, and thermal energy storage. Reproduced with permission from (Fernandes et al., 2025).

Hybrid systems are enabling smart materials that self-heal, mimic motion, or adapt to their environment. Examples include hybrid actuators used as artificial muscles in soft robotics and self-repairing materials triggered by heat or light—vital for aerospace, biomedical, and wearable tech. Combining metal oxides with conductive polymers has also led to advanced gas sensors and responsive actuators. As research progresses, such hybrids are set to transform sectors from healthcare and robotics to smart manufacturing and AI.

2.5 Challenges in Hybrid Material Development

Hybrid materials, though increasingly prominent across scientific and industrial fields, face hurdles like poor interfacial compatibility, complex fabrication, limited durability, and high costs. Tackling these challenges is vital for scaling up from lab research to practical applications.

2.5.1 Interfacial Compatibility

A major challenge in hybrid material design is ensuring strong bonding between organic and inorganic components. Poor interfacial compatibility can cause phase separation, weaken mechanical properties, and lead to unstable behavior during use. In metal–polymer composites, thermal or chemical mismatches can cause delamination or microcracks. Likewise, weak bonding in UHTCMCs may reduce thermal resistance and structural strength.

Advanced modeling, surface treatment, and interface engineering are key to resolving compatibility issues and ensuring long-term material stability.

2.5.2 Scalability and Production Costs

Scaling up high-quality hybrid material synthesis remains difficult due to complex, time-consuming methods like CVI and PIP, which demand strict processing conditions and drive up costs. High costs from using rare materials like noble metals and specialized ligands hinder large-scale hybrid material deployment. To address this, industries are turning to bio-based, low-cost alternatives, and some startups are shifting toward scalable markets like automotive and furniture.

2.5.3 Stability and Long-Term Durability

Maintaining long-term chemical, mechanical, and thermal stability is a key challenge for hybrid materials, particularly under fluctuating conditions like humidity, UV exposure, temperature swings, or mechanical stress.

Some hybrid systems perform well short term, but sustaining performance in harsh conditions is difficult. Organic components often degrade faster than inorganic ones, risking loss of function or structural failure.

Extending material life and recyclability is crucial for energy storage, structural, and catalytic applications. This calls for ongoing research into self-healing, encapsulation, and durable surface passivation strategies.

Unlocking the full potential of hybrid materials requires multidisciplinary innovation across chemistry, materials science, and engineering. Advancing scalable synthesis, robust interface design, and sustainable components is key to enabling broader adoption and real-world impact.

2.6 Future Prospects

The future of hybrid materials rests in their capacity to tackle complex global challenges through smart design, sustainability, and adaptive functionality. Breakthroughs in nanotech, green chemistry, and modeling are enabling next-gen systems that are more intelligent, eco-friendly, and responsive than ever before.

2.6.1 Nanostructured Hybrid Materials

Nanostructured hybrids are a leading focus in hybrid material research, offering atomic-level control over properties like surface area, morphology, and electronic or optical behavior. These features make them ideal for applications in sensing, biomedicine, electronics, and photocatalysis. Recent advances in hybrid nanocomposites have improved electrochemical sensor sensitivity and enabled targeted drug delivery and regenerative therapies. Their nanoscale design enhances cell interaction, biodegradability, and multifunctionality—key for personalized healthcare.

2.6.2 Sustainable Hybrid Materials

With rising demand for eco-friendly solutions, sustainability is now central to hybrid material research. Scientists are turning to renewable feedstocks and biological systems for greener synthesis methods.

A novel strategy uses fungal mycelium to grow bio-hybrid composites on lingo-cellulosic agricultural waste. These lightweight, biodegradable, and fire-resistant materials are scalable, moldable, and ideal for sustainable packaging, construction, and interior design.

Hybrid materials are transforming sustainable architecture through smart energy systems like thermoelectric coatings, phase change materials, and solar-integrated building elements.

2.6.3 Self-Healing and Smart Hybrid Materials

Next-gen hybrid materials are set to be smarter and more adaptive—featuring self-healing, sensing, and stimulus-responsive abilities. Such innovations are vital for demanding fields like aerospace, robotics, and healthcare.

Hybrid systems are enabling self-repairing electronics that restore conductivity or structure after damage. In construction, recyclable asphalt with self-healing agents helps fix micro cracks, boosting road durability and cutting maintenance costs.

Smart hybrid materials that respond to temperature, light, or stress are enabling responsive surfaces, artificial muscles, and wearable sensors. They extend device longevity and improve interaction with users and environments.

As fields converge, future hybrid materials will be multifunctional, eco-friendly, and digitally connected—driving the next wave of innovation.

2.7 Conclusion

Advanced hybrid materials lead modern materials science by uniting the strengths of organic and inorganic components into multifunctional, high-performance systems. They're driving breakthroughs in energy storage, catalysis, environmental cleanup, and sensing—where conventional materials often fall short in adaptability, sustainability, and precision.

Through nanoscale design, researchers are advancing hybrid materials with custom properties like high conductivity, adjustable porosity, resilience, and responsiveness. These traits are especially valuable for flexible electronics, medical devices, and sustainable infrastructure.

Hybrid materials hold great promise, but challenges like interfacial compatibility, long-term stability, cost, and scalability remain. Overcoming these hurdles through interdisciplinary research is crucial for their commercial success. Hybrid materials are set to address key 21st-century challenges—from climate change to sustainable manufacturing and personalized medicine. Advances in synthesis, computational tools, and AI-driven discovery will speed up innovation and impact across sectors. As science advances and technology evolves, hybrid materials will reshape material capabilities and drive smarter, greener, and more resilient solutions for global progress.

References

Ali, N., Khan, F., Khan, I., Kareem, A., Rahman, S. et al. 2024. Robust polymer hybrid and assembly materials from structure tailoring to efficient catalytic remediation of emerging pollutants. *Chemosphere*, 142408.

Ali, S.W., Bairagi, S. and Islam, S.U. (eds.). 2024. *Hybrid Materials for Piezoelectric Energy Harvesting and Conversion*. Wiley.

Badoni, A., Thakur, S., Vijayan, N., Swart, H.C., Bechelany, M. et al. 2025. Recent progress in understanding the role of graphene oxide, TiO_2 and graphene oxide-TiO_2 nanocomposites as multidisciplinary photocatalysts in energy and environmental applications. *Catal. Sci. Technol.*, 15: 1702–1770.

Chongdar, S., Bhattacharjee, S., Bhanja, P. and Bhaumik, A. 2022. Porous organic–inorganic hybrid materials for catalysis, energy and environmental applications. *Chemical Communications*, 58(21): 3429–3460.

Chuhadiya, S., Suthar, D., Patel, S.L. and Dhaka, M.S. 2021. Metal organic frameworks as hybrid porous materials for energy storage and conversion devices: A review. *Coordination Chemistry Reviews*, 446: 214115.

Dou, Q., Wu, N., Yuan, H., Shin, K.H., Tang, Y. et al. 2021. Emerging trends in anion storage materials for the capacitive and hybrid energy storage and beyond. *Chemical Society Reviews*, 50(12): 6734–6789.

Fagiolari, L., Sampò, M., Lamberti, A., Amici, J., Francia, C. et al. 2022. Integrated energy conversion and storage devices: Interfacing solar cells, batteries and supercapacitors. *Energy Storage Materials*, 51: 400–434.

Fernandes, L.C., Correia, D.M., Costa, C.M. and Lanceros-Mendez, S. 2025. Recent advances in ionic liquid-based hybrid materials for electroactive soft actuator applications. *Macromolecular Materials and Engineering*, 310(2): 2400279.

Fu, Y., Liu, T., Wang, H., Wang, Z., Hou, L. et al. 2024. Applications of nanomaterial technology in biosensing. *Journal of Science: Advanced Materials and Devices*, 100694.

Gao, X., Wu, H., Su, C., Lu, C., Dai, Y. et al. 2023. Recent advances in carbon-based nanomaterials for multivalent-ion hybrid capacitors: A review. *Energy & Environmental Science*, 16(4): 1364–1383.

Gassara, M., Hchicha, K., Mhadhbi, N., Ahmed, A.B., Guesmi, A. et al. 2025. Insights into Co (II)-based hybrid materials for photocatalytic and antioxidant applications. *Journal of Molecular Structure*, 1319: 139458.

Gideon, O., Samuel, H.S. and Okino, I.A. 2024. Biocompatible materials for next-generation biosensors. *Discover Chemistry*, 1(1): 34.

Gleason, K.K. 2024. Designing organic and hybrid surfaces and devices with initiated chemical vapor deposition (iCVD). *Advanced Materials*, 36(8): 2306665.

Godja, N.C. and Munteanu, F.D. 2024. Hybrid nanomaterials: A brief overview of versatile solutions for sensor technology in healthcare and environmental applications. *Biosensors*, 14(2): 67.

Gomez-Romero, P., Pokhriyal, A., Rueda-García, D., Bengoa, L.N., González-Gil, R.M. et al. 2023. Hybrid materials: A meta review. *Chemistry of Materials*, 36(1): 8–27.

Grisolia, A., Dell'Olio, G., Spadafora, A., De Santo, M., Morelli, C. et al. 2023. Hybrid polymer-silica nanostructured materials for environmental remediation. *Molecules*, 28(13): 5105.

Ivanova, V., Şenocak, A., Klyamer, D., Demirbas, E., Makhseed, S. et al. 2024. 3D hybrid materials based on zinc and cobalt phthalocyanine derivatives bearing terminal ethynyl substituents as active layers for chemiresistive gas sensors. *Sensors and Actuators B: Chemical*, 398: 134733.

Jyoti, J., Gupta, T.K., Singh, B.P., Sandhu, M., Tripathi, S.K. et al. 2022. Recent advancement in three dimensional graphene-carbon nanotubes hybrid materials for energy storage and conversion applications. *Journal of Energy Storage*, 50: 104235.

Liu, S., Kang, L., Zhang, J., Jun, S.C., Yamauchi, Y. et al. 2021. Carbonaceous anode materials for non-aqueous sodium-and potassium-ion hybrid capacitors. *ACS Energy Letters*, 6(11): 4127–4154.

Merghes, P., Ilia, G., Maranescu, B., Varan, N., Simulescu, V. et al. 2024. The sol–gel process, a green method used to obtain hybrid materials containing phosphorus and zirconium. *Gels*, 10(10): 656.

Nawaz, F., Ali, M., Ahmad, S., Yong, Y., Rahman, S. et al. 2024. Carbon based nanocomposites, surface functionalization as a promising material for VOCs (volatile organic compounds) treatment. *Chemosphere*, 143014.

Ndlwana, L., Raleie, N., Dimpe, K.M., Ogutu, H.F., Oseghe, E.O. et al. 2021. Sustainable hydrothermal and solvothermal synthesis of advanced carbon materials in multidimensional applications: A review. *Materials*, 14(17): 5094.

Nirwan, V.P., Kowalczyk, T., Bar, J., Buzgo, M., Filová, E. et al. 2022. Advances in electrospun hybrid nanofibers for biomedical applications. *Nanomaterials*, 12(11): 1829.

Porcu, S., Secci, F. and Ricci, P.C. 2022. Advances in hybrid composites for photocatalytic applications: A review. *Molecules*, 27(20): 6828.

Potbhare, A.K., Aziz, S.T., Ayyub, M.M., Kahate, A., Madankar, R. et al. 2024. Bioinspired graphene-based metal oxide nanocomposites for photocatalytic and electrochemical performances: An updated review. *Nanoscale Advances*, 6(10): 2539–2568.

Prasannan, A., Udomsin, J., Tsai, H.C., Wang, C.F., Lai, J.Y. et al. 2020. Robust underwater superoleophobic membranes with bio-inspired carrageenan/laponite multilayers for the effective removal of emulsions, metal ions, and organic dyes from wastewater. *Chem. Eng. J.*, 391: 123585.

Roy, S., Miller, M., Warnan, J., Leung, J.J., Sahm, C.D. et al. 2021. Electrocatalytic and solar-driven reduction of aqueous CO_2 with molecular cobalt phthalocyanine–metal oxide hybrid materials. *ACS Catalysis*, 11(3): 1868–1876.

Song, J., Vikulina, A.S., Parakhonskiy, B.V. and Skirtach, A.G. 2023. Hierarchy of hybrid materials. Part-II: The place of organics-on-inorganics in it, their composition and applications. *Frontiers in Chemistry*, 11: 1078840.

Sun, B., Xu, G., Ji, X., Yang, Z., Guan, C. et al. 2024. A strain-resistant flexible thermistor sensor array based on CNT/MXene hybrid materials for lithium-ion battery and human temperature monitoring. *Sensors and Actuators A: Physical*, 368: 115059.

Sun, G., Chen, D., Zhu, G. and Li, Q. 2022. Lightweight hybrid materials and structures for energy absorption: A state-of-the-art review and outlook. *Thin-Walled Structures*, 172: 108760.

Thakur, A. and Kumar, A. 2024. Challenges and opportunities in the development of nano-hybrid smart coatings. *Nano-Hybrid Smart Coatings: Advancements in Industrial Efficiency and Corrosion Resistance*, 353–384.

Xing, F., Bi, Z., Su, F., Liu, F., Wu, Z.S. et al. 2022. Unraveling the design principles of battery-supercapacitor hybrid devices: From fundamental mechanisms to microstructure engineering and challenging perspectives. *Advanced Energy Materials*, 12(26): 2200594.

Xu, K., Zhou, Z.N., Han, X.B., Yang, Y.W., Zhang, W. et al. 2024. Shape shifting and locking in mechanically responsive organic-inorganic hybrid materials for thermoelastic actuators. *Angewandte Chemie*, 136(33): e202408247.

Yan, Huchuan 2023. Metal-carbon hybrid materials induced persulfate activation: Application, mechanism, and tunable reaction pathways. *Water Research*, 234, 119808.

Zeb, H. and Riaz, A. 2022. Introduction to organic–inorganic nanohybrids. In *hybrid Nanomaterials: Biomedical, Environmental and Energy Applications* (pp. 1–27). Singapore: Springer Nature Singapore.

Zhou, C., Xu, L.J., Lee, S., Lin, H., Ma, B. et al. 2021. Recent advances in luminescent zero-dimensional organic metal halide hybrids. *Advanced Optical Materials*, 9(18): 2001766.

Zhou, K., Qi, B., Liu, Z., Wang, X., Sun, Y. et al. 2024. Advanced organic–inorganic hybrid materials for optoelectronic applications. *Advanced Functional Materials*, 34(52): 2411671.

Zhu, K., Cheng, Z., Rangan, S., Cotlet, M., Du, J. et al. 2021. A new type of hybrid copper iodide as nontoxic and ultrastable LED emissive layer material. *ACS Energy Letters*, 6(7): 2565–2574.

Zhu, N., Zhang, K., Wu, F., Bai, Y. and Wu, C. 2021. Ionic liquid-based electrolytes for aluminum/magnesium/sodium-ion batteries. *Energy Mater Adv.*, 2021. DOI:10.34133/2021/9204217.

Zhu, Y., Zhang, S., Qiu, X., Hao, Q., Wu, Y. et al. 2024. Graphdiyne/metal oxide hybrid materials for efficient energy and environmental catalysis. *Chem. Sci.*, 15: 5061–5081.

CHAPTER

3 Green Nanoparticles for a Sustainable Future

Synthesis, Potential and Applications

Suman Dasmodak, *Sudip Paramanik*, *Sagarika Mukherjee* and *Manas Paramanik**

Nanotechnology is rapidly emerging across fields like engineering, medicine, and pharmacology due to its wide-ranging applications. Green nanotech, in particular, stands out for offering eco-friendly, low-toxicity solutions. Nanoparticles under 100 nm exhibit high surface-area-to-volume ratios, giving them distinct physical and chemical properties unlike their bulk counterparts. Green synthesis of NPs nanoparticles relies on eco-friendly solvents, biogenic reducers, stabilizers, and capping agents for stability. Plants, microbes, fungi, and biopolymers act as reducing agents, while metals like Cu, Au, Ag, Pd, Pt, TiO_2, and ZnO serve as base materials. Green-synthesized hybrid nanomaterials are characterized using UV-Vis, TGA, GC-MS, XRD, and DLS for physicochemical analysis, while SEM, AFM, and TEM confirm their structure. These NPs show strong potential in clinical applications like drug delivery, antimicrobial therapy, and vector control, as well as in environmental cleanup. The chapter emphasizes sustainable synthesis methods and their broad industrial relevance.

Epidemiology, Vector Biology and Environmental Monitoring Research Units, Entomology Laboratory, Department of Animal Science, Kazi Nazrul University, Asansol, India.
* Corresponding author: manasp74@gmail.com
ORCID: http://orcid.org/0000-0002-0312-5082

DOI: 10.1201/9781003607359-3

3.1 Introduction

Nanotechnology, first coined by Prof. Norio Taniguchi of Tokyo University of Science, involves manipulating materials at the atomic or molecular scale—typically under 100 nm (Mansoori, 2005). This nanoscale enables unique quantum and optical effects, such as electron trapping, enhancing detection and reactivity (Hashem et al., 2022; Rautela and Rani, 2019). Their high surface-to-volume ratio also improves interaction with biological systems, boosting performance across various fields (Paul et al., 2014; Paramanik et al., 2023a). Conventional chemical and physical nanoparticle synthesis often produces harmful by-products, posing environmental risks. In contrast, green methods have gained traction for yielding nanoparticles with superior properties—such as Surface Plasmon Resonance and enhanced light scattering—compared to bulk metals. These eco-friendly metal/metal oxide NPs are now widely explored in catalysis, biomedicine, agriculture, and energy sectors (Mohanta et al., 2020; Singh et al., 2021; Abishek et al., 2024). Sustainable nanoscience, grounded in green chemistry, promotes safe, eco-friendly, and cost-effective nanomaterial synthesis (Bai et al., 2018). It uses plant parts, microbes, fungi, enzymes, and thermal methods as alternatives to toxic chemicals (Kahrilas et al., 2014). Plant extracts rich in saponins, terpenoids, phenols, and alkaloids aid in reducing and stabilizing metal ions, forming nanoparticles with varied shapes, sizes, and durability (Mohanta et al., 2020; Singh et al., 2023). Optimal temperature, pH, and pressure are crucial for efficient synthesis (Patra and Baek, 2014). Synthesized nanoparticles are characterized using various techniques to assess their size, shape, stability, crystallinity, surface charge, and composition. UV-Vis spectroscopy monitors nanoparticle formation via surface plasmon resonance peaks. TGA reveals thermal stability and composition by tracking weight loss under heat. XRD identifies crystal structure and purity, while MS determines molecular weight and elemental makeup. Dynamic light scattering (DLS) determines nanoparticle hydrodynamic diameter and zeta potential, indicating dispersion and stability. Microscopic techniques reveal structural details: SEM captures surface morphology at the nanoscale, TEM uncovers internal structure and size distribution, and AFM maps topography and mechanical properties in 3D (Akhtar et al., 2018; Sneddon et al., 2016; Sharma et al., 2022).

This chapter highlights eco-friendly methods for nanoparticle synthesis, focusing on using plants, microbes, and fungi as natural stabilizers, capping, and reducing agents in place of toxic chemicals. It also outlines the wide-ranging applications of green nanoparticles in environmental cleanup, healthcare, and agriculture, while exploring their future potential in sustainable nanotechnology.

3.2 Green Synthesis of Nanoparticles

Nanoparticles can be synthesized via physical, chemical, biological, and green methods. Chemical approaches—like reduction, electrochemical, irradiation-mediated, and pyrolysis techniques—use metal precursors, chemical reducers (e.g., alcohol, ascorbic acid, hydrazine, borohydride), and stabilizers (Wei et al., 2015; Sotiriou and Pratsinis, 2010). Physical processes are extremely efficient and do not involve harmful chemicals. Physical methods, including ball milling, vapor condensation, DC magnetron sputtering, and arc discharge, offer precise, uniform nanoparticles but require high energy (Wei et al., 2015; Asanithi et al., 2012). To overcome such drawbacks, researchers are increasingly focusing on eco-friendly green and biosynthetic alternatives. Green nanoparticle synthesis follows the 12 principles of green chemistry, emphasizing non-toxic reagents, renewable resources, and biodegradable by-products (Madani et al., 2022). Like chemical methods, it involves three steps: solvent selection, reduction, and stabilization—using plant-based phytochemicals as eco-friendly agents for size and shape control. These phytochemicals are abundant, biocompatible, and effective. Alternatives like fungi, algae, and microbes have also been explored, though they demand sterile conditions and are resource-intensive (Narayanan and Sakthivel, 2011; Raveendran et al., 2003; Prakash et al., 2013). Nanoparticle morphology depends on extract type and concentration—e.g., Rhizopus oryzae yields flower-like forms at high extract levels, but spherical or triangular ones at lower levels (Das et al., 2010). Green nanoparticle synthesis involves three main stages: (1) preparing a biogenic aqueous extract as the reaction medium, (2) reducing metal ions to form nucleation centers and initiate nanoparticle growth, and (3) stabilizing the particles to enhance their properties and prevent aggregation (Makarov et al., 2014; Fig. 3.1).

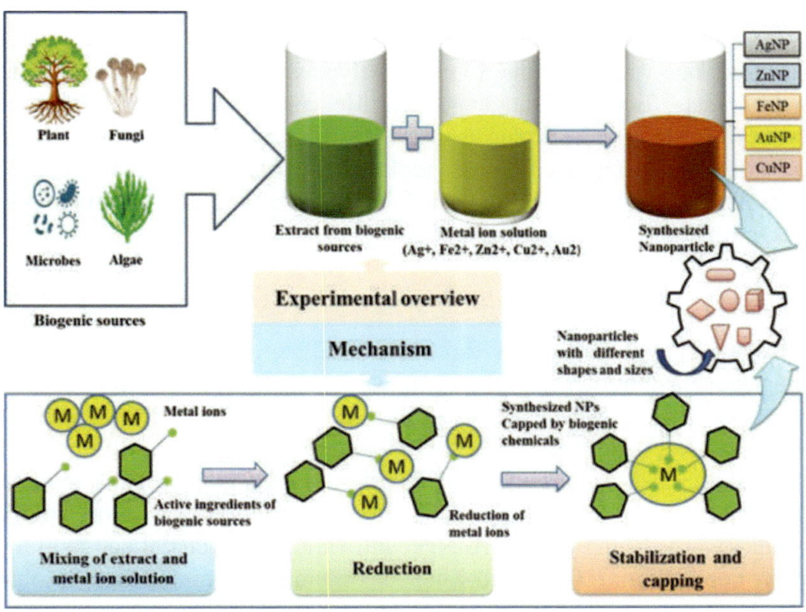

Figure 3.1: Schematic of green synthesis process.

3.2.1 Green Synthesis of Silver Nanoparticles (AgNPs)

Silver (Ag) has been valued for its antimicrobial properties for over 5,000 years, with ancient civilizations using it for food preservation and utensils (Srikar et al., 2016). Its strong bactericidal action led to diverse uses—in medicines like silver nitrate and sulfadiazine for wound care, in household appliances, and food packaging. This long-standing efficacy inspired the development of silver nanoparticles (AgNPs), which exhibit enhanced biological activities including antimicrobial, anti-larval, anti-inflammatory, and anti-tumour effects (Bruna et al., 2021; Algotiml et al., 2022; Samiei et al., 2024; Aravinth et al., 2024). Several studies support the green synthesis of AgNPs using various biogenic sources. Sangameshwaran et al. (2024) used Senna alexandrina leaf extract to produce spherical AgNPs (20–35 nm), confirmed by UV-Vis (440 nm) and FE-SEM, showing strong larvicidal activity. Shah et al. (2021) synthesized AgNPs (26–34 nm) from Plantago lanceolata extract, verified by EDX (54.87% Ag), with notable antibacterial effects. Majumder and Kar (2024) generated spherical, crystalline AgNPs (20–50 nm) from Abelmoschus esculentus calyx, demonstrating significant anthelminthic action. Rautela and Rani (2019) used Tectona grandis seed extract to yield AgNPs (10–30 nm), stabilized by phytochemicals, with FTIR confirming active functional groups. These AgNPs disrupted bacterial membranes effectively. Baran et al. (2023) synthesized stable AgNPs from Allium cepa peel extract via $AgNO_3$ reduction, showing SPR color change, strong thermal stability, high zeta potential, antioxidant properties, and novel anticholinesterase activity relevant to Alzheimer's treatment.

Castañeda-Aude et al. (2023) synthesized ultra-small, monodispersed AgNPs (avg. 2.42 nm) using citrus peel extracts, effective against Staphylococcus aureus. Their small size was likely due to citrus polyphenols. Klein et al. (2023) used Searsia crenata, Berzelia lanuginosa, and Helichrysum cymosum to produce AgNPs (16–72 nm) with varying shapes and strong antifungal activity. Sana and Dogiparthi (2018) employed Givotia moluccana extract, yielding stable AgNPs (30–40 nm) with broad-spectrum antibacterial effects. Table 3.1 summarizes agents, morphologies, and applications.

3.2.2 Green Synthesis of Gold Nanoparticles (AuNPs)

Like AgNPs, phyto-compounds can reduce Au^{3+} to Au^0 and stabilize AuNPs, making green synthesis using plant extracts a preferred, eco-friendly approach (Ahmed and Ikram, 2016). Yayintas et al. (2024) used Nasturtium officinale to produce ~56.4 nm AuNPs, which showed cytotoxicity against A549 lung cells. Botteon et al. (2021) synthesized AuNPs using Brazilian red propolis extracts—hexane (rich in benzophenones) and ethyl acetate (rich in isoflavonoids/flavonoids)—producing distinct AuNPs with varied antibacterial and antifungal activity. Fouda et al. (2022) used Zingiber officinale rhizome extract to obtain stable globular AuNPs (5–53 nm, zeta potential –28.8 mV) with strong

Table 3.1: Some green synthesized silver nanoparticles (AgNPs) with their synthesizing agent, morphology, size, and application.

Natural synthesizing agent	Morphology	Size (nm)	Applications	References
Tectona grandis	Spherical, Oval	10–30	Antimicrobial	Rautela and Rani, 2019
Azolla pinnata	Spherical	10–20	Antilarval	Wilson et al., 2023
Catharanthus roseus	Cubic	35–55	Antiplasmodial	Ponarulselvam et al., 2012
Coffea arabica	Spherical, ellipsoidal	20–30	Antibacterial	Dhand et al., 2016
Ocimum gratissimum	Spherical	69.0±5	Antimicrobial, Anticancer	Alex et al., 2024
Ocimum tenuiflorum	Spherical	46.9±9	Antimicrobial, Anticancer	Alex et al., 2024
Ocimum americanum	Spherical	58.5±18.7	Antimicrobial, Anticancer	Alex et al., 2024
Curcuma longa	Spherical	70	Antibacterial, Antioxidant	Din et al., 2024
Moringa oleifera	Spherical	80	Antibacterial, Antioxidant	Din et al., 2024
Pteris tripartita Sw.	Hexagonal, spherical, rod shaped	32	Antioxidant, Anti-inflammatory, Antibacterial	Baskaran et al., 2016
Salvia spinosa	Spherical	5.13	Antibacterial	Pirtarighat et al., 2019
Camellia sinensis	Spherical	17.6	Antibacterial	Nakhjavani et al., 2017
Aspergillus sydowii	Spherical	1–24	Antifungal, Antiproliferative	Wang et al., 2021
Abelmoschus esculentus	Spherical	20–50	Antihelminthic	Majumdar and Kar, 2024
Couroupita guianensis	Spherical	5-30	Antibacterial	Rajesh Kumar et al., 2016
Caulerpa racemose	Spherical, triangle	5-25	Antibacterial	Aboelfetoh et al., 2017
Holorrhena antidysenterica	Spherical	32	Antilarval	Kumar et al., 2018
Portieria hornemanni	Spherical	70–75	Antibacterial	Fatima et al., 2020
Givotia moluccana	Spherical	30–40	Antimicrobial	Sana and Dogiparthi, 2018

antimicrobial and selective anticancer properties. Majumdar et al. (2016) demonstrated green synthesis of AuNPs using Mimusops elengi bark extract rich in saponins and tannins, which reduced and stabilized AuNPs at room temperature. These particles catalyzed the conversion of 3- and 4-nitrophenol to aminophenol. Honey has also served as a reducing and stabilizing agent due to its fructose and vitamin C content, yielding ~15 nm spherical AuNPs with crystalline structure (Philip, 2009). Wang et al. (2021) used Viola betonicifolia extract, where flavonoids and phenolics reduced Au^{3+} via redox reactions, while other metabolites stabilized the nanoparticles. Singh and Srivastava (2015) synthesized AuNPs from black cardamom, where 1,8-cineole was oxidized to 2-oxo-1,8-cineole. The particle size varied with pH 20–40 nm at pH 7, but larger, irregular, or flower-shaped forms at lower or higher pH. Soliman et al. (2024) biosynthesized AuNPs from Fusarium pseudonygamai, producing 8–60 nm particles with significant antibacterial activity. Table 3.2 summarizes synthesizing agents, nanoparticle properties, and applications.

Table 3.2: Some green synthesized gold nanoparticles (AuNPs) with their synthesizing agent, morphology, size, and application.

Natural synthesizing agent	**Morphology**	**Size (nm)**	**Applications**	**References**
Turbinaria conoides	Tri-spherical, triangle, pseudo-spherical	6–10	Antibacterial	Rajeshkumar et al., 2013
Lippie citriodora	Triangular, hexagonal	36	Imaging	Elia et al., 2014
Justicia adhatoda	Spherical	13–57	Catalyst	Latha et al., 2018
Elaise guineensis	Spherical, triangular, pentagonal	27.89	–	Ahmad et al., 2016
Pelargonium graveolens	Spherical	45	Drug delivery, Imaging	Elia et al., 2014
Anacardium occidentale	Spherical	10–30	Antimicrobial, Anticancer	Sunderam et al., 2019
Desmodium gangeticum	Spherical	16±4	Antioxidant	Ghosh et al., 2020
Allium cepa L.	Spherical	6.08–54.20	Antipathogenic, Antioxidant	İpek et al., 2024
Croton sparsiforus	Spherical	16.6–17	UV-protection, antibacterial, and anticancer agents	Boomi et al., 2020
Cinnamomum zeylanicum	Spherical	25	Optoelectronic, biomedical	Smitha et al., 2009
Mangifera indica L.	Kidney-shaped	30–50	Drug delivery	Patra et al., 2018
Litsea cubeba	Spherical	8–18	Catalytic reduction of 4-nitrophenol	Doan et al., 2020
Macrotyloma uniflorum	Spherical	14–17	Pharmaceuticals, biomedical	Aromal et al., 2012
Couroupita guianensis	Spherical, triangular and hexagonal	26±11	Antioxidant	Sathishkumar et al., 2016

3.2.3 Green Synthesis of Copper Nanoparticles (CuNPs)

Researchers have successfully synthesized CuNPs using green methods with plant and biological extracts. Ocimum sanctum leaf extract produced CuNPs (~13.1 nm), confirmed via FTIR, TEM, and XRD, and showed antibacterial and anticancer effects (Ashokkumar et al., 2024). Rubia cordifolia-based CuNPs (~50.7 nm) displayed mixed Cu, Cu_2O, and CuO phases with strong antimicrobial activity (Vinothkanna et al., 2023). Spondias mombin extract yielded CuO NPs sized 10–75 nm (Abishek and Kachhap, 2025). Fish scales of Labeo rohita also served as green reducers and stabilizers, generating spherical CuNPs (25–37 nm) with characteristic SPR at 580 nm and fcc crystal structure (Sinha and Ahmaruzzaman, 2015). Table 3.3 summarizes synthesizing agents, sizes, shapes, and applications.

Table 3.3: Some green synthesized copper nanoparticles (CuNPs) with their synthesizing agent, morphology, size, and application.

Natural synthesizing agent	Morphology	Size (nm)	Applications	References
Hagenia abyssinica	Spherical, hexagonal, triangular, cylindrical	34.76	Antimicrobial	Murthy et al., 2020
Coccinia grandis	Spherical	40–50	Catalyst	Devi et al., 2018
Native cyclodextrins	Spherical	2–33	–	Suárez-Cerda, 2017
Cocculus hirsutus	Sheet like	63.46	Antidiabetic, Antioxidant, Antimicrobial	Ameena et al., 2022
Achillea fragrantissima	Spherical	4.61–6.97	Larvicidal, Adulticidal	Abd El-Halim et al., 2025
Fortunella margarita	Spherical	51.2–56.6	Drug delivery	Amjad et al., 2021
Orobanche aegyptiaca	Spherical	< 50	Nematicidal activity	Akhter et al., 2020
Azadirachta indica	Cubical	28–30	Antibacterial	Sharma et al., 2018

3.2.4 Green Synthesis of Additional Metal Oxide and Metal Nanoparticles

Beyond AuNPs, AgNPs, and CuNPs, other metal-based nanoparticles like ZnO, MgO, Fe_2O_3, TiO_2, Pt, Fe, and Pd show promising nanoscale features for applications in plastics, pharmaceuticals, and nano-therapeutics. For instance, ZnONPs synthesized using Brassica oleracea leaf extract showed a wurtzite-hexagonal structure and 52 nm particle size with strong larvicidal activity (Manojkumar et al., 2023). ZnONPs from Cucurbita seed extract ranged from 45–65 nm, with rectangular, hexagonal, and rod-like shapes due to seed phytochemicals (Velsankar et al., 2019). Dinga et al. (2022) used *Melia azedarach* extract to green-synthesize MgO nanoparticles, confirmed by XRD. SEM and TEM revealed internal and surface morphology, with TEM showing that higher extract concentrations yielded smaller particles. Separately, Pterolobium hexapetalum extract produced Fe_2O_3NPs (20–28 nm), characterized by a 290 nm UV-Vis peak and FTIR bands (C–C, C–H, C=C, O–H, C–X), exhibiting strong larvicidal activity against Culex larvae (Loganathan et al., 2023). Several researchers developed nanoparticles from TiO_2 nanoparticles (TiO_2NPs) are valued for their stability and multifunctional optical and electronic properties. Balaraman et al. (2022) used Sargassum myriocystum seaweed extract to biosynthesize TiO_2NPs (50–90 nm) with cubic to spherical shapes and crystalline structure. The particles exhibited a negative surface charge. Table 3.4 lists other green-synthesized metal/metal oxide nanoparticles, their morphology, size, and applications.

Even with natural stabilizers, nanoparticles may agglomerate post-synthesis. To enhance stability, methods such as ultrasonication, mechanical stirring, steric and electrostatic stabilization, or combined deagglomeration are employed (Khairunnisa et al., 2021).

3.3 Potential and Application of Green Synthesized Nanoparticles

Synthetic chemicals, pose serious risks to human and ecological health (Dasmodak et al., 2024a; Paramanik et al., 2024a), prompting a shift toward green alternatives (Paramanik et al., 2022). As a result, global research on phytochemicals is expanding (Chowdhury et al., 2015; Paramanik, 2018; Paramanik et al., 2023b, 2024b; Dasmodak et al., 2024b). Green-synthesized nanoparticles are gaining traction as sustainable solutions across multiple fields (Fig. 3.2).

3.3.1 Biomedical Applications

Green-synthesized nanoparticles offer key advantages—redox activity, high surface-area-to-volume ratio, structural stability, and strong biocompatibility—making them valuable in biomedicine (Andreescu et al., 2012; McNamara and Tofail, 2017). Metal and metal oxide NPs are widely explored for biosensing, bioimaging, diagnostics, and stem cell tracking. AgNPs, in particular, excel in antimicrobial, anticancer, and regenerative applications, including drug delivery and tissue scaffolding. They've shown potent bactericidal action against pathogens like *V. cholerae*, *P. aeruginosa*, *S. typhus*, *S. enterica*, *B. cereus*, and *B. subtilis* (Azeem et al., 2022; Vijayaram et al., 2024). AgNP-coated masks have shown strong efficacy against *E. coli* and *S. aureus*, making them ideal for hospital infection control. AgNP-based disinfectants effectively decontaminate surgical

Table 3.4: Some additional green synthesized nanoparticles with their synthesizing agent, morphology, size, and application.

Natural synthesizing agent	Metal/ Metal-oxide	Morphology	Size (nm)	Applications	References
Brassica oleracea	ZnO	Hexagonal and wurtzite	52	Antilarval	Manojkumar et al., 2023
Moringa oleifera	ZnO	Spherical	12.27–30.51	–	Matinise et al., 2017
Citrus lemon	ZnO	Spherical	21.5	Dye degradation	Davar et al., 2015
Ambrosia ambrosioides	ZnO	Irregular	5–56	Dye degradation	Medina-Acosta et al., 2024
Syzygium aqueum	Pd	Irregular	5–20	Catalysis in the coupling reaction	Manjare and Chaudhari, 2020
Dioscorea bulbifera	Pd-Pt	Spherical	10–25	Anticancer and Antioxidant	Ghosh et al., 2015
Eucommia ulmoides	Pd	Spherical	12.6	Catalyst	Duan et al., 2015
Santalum album	Pd	Spherical	10–40	Antibacterial and catalytic activity	Sharmila et al., 2017
Cochlospermum gossypium	Pd	Spherical	6.5	Catalyst	Rastogi et al., 2015
Tea polyphenol	Pt	Flower shape	30–60	Cancer treatment	Alshatwi et al., 2015
Padina gymnospora	Pt	Spherical	< 35	Catalyst	Shiny et al., 2014
Quercus dalechampii	Pt	Spherical	58.5±7.6	Antioxidant and antimicrobial	Coman et al., 2024
Quercus frainetto	Pt	Spherical	41.6±5.4	Antioxidant and antimicrobial	Coman et al., 2024
Quercus petraea	Pt	Spherical	41±5.3	Antioxidant and antimicrobial	Coman et al., 2024
Tinospora cordifolia	$Mg(OH)_2$	Spherical	27.71	Antibacterial	Rajkumar et al., 2024
Penicillium crustosum	MgO	Spherical	8–35	Growth and development	Amin et al., 2024
Rosa floribunda charisma	MgO	Polyhedral	35.25–55.14	Antioxidant, antiaging, antibiofilm	Younis et al., 2021
Hyphaene thebaica	MgO	Quasi-spherical	32.6±5	Catalyst	Muhaymin et al., 2024
Mentha arvensis	TiO_2	Spherical	20–70	Antimicrobial	Ahmad et al., 2020
Psidium guajava	Fe_2O_3	Spherical	29–41	Antifungal	Anwaar et al., 2024

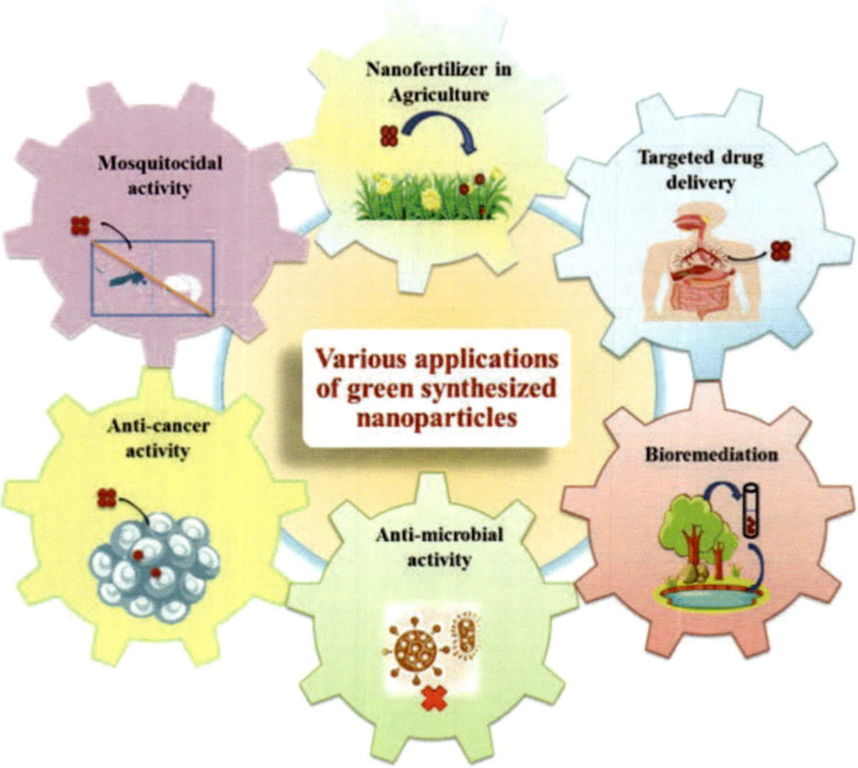

Figure 3.2: Diverse applications of green-synthesized nanoparticles in different domains.

masks, while coated catheters inhibit bacterial and fungal biofilms. Animal studies reported no non-specific toxicity, supporting their long-term antimicrobial use in medicine (Li et al., 2006; Roe et al., 2008; Hiragond et al., 2018). Fe_2O_3 nanoparticles are widely used in biomedicine for MRI contrast enhancement, drug delivery, thermal tumor therapy, and bio-separation due to their magnetic properties (Gupta and Gupta, 2005). AuNPs are chemically and thermally reactive, biocompatible, and ideal for drug delivery, diagnostics, gene therapy, and bioimaging. When combined with photosensitizers, they enable photo-activated antibacterial treatment (Managa et al., 2014; Shivaramakrishnan et al., 2017). Their strong light-scattering cross-section—3–5× greater than fluorophores—makes them highly effective for dark-field microscopy (Aminabad et al., 2019; Gupta and Gupta, 2005).

Green-synthesized nanoparticles are emerging as efficient tools for advanced drug delivery, diagnostics, and treatment of serious diseases. Their phytochemical components can suppress cancer progression and trigger apoptosis (Wani et al., 2023). Superparamagnetic Fe_2O_3NPs are especially promising for targeted drug delivery due to high drug-loading capacity, though surface modifications are often needed for controlled release (Ulbrich et al., 2016).

3.3.2 Environmental Applications

Green nanomaterials play a critical role in environmental remediation by removing toxic pollutants and treating wastewater. Industries like textiles release large volumes of organic dyes that cloud water, block sunlight, and harm aquatic ecosystems (Carmen and Daniela, 2012; Singh et al., 2018). While conventional methods like oxidation and electrochemical treatment can be effective, they often impose environmental burdens (Crini and Lichtfouse, 2018). Metal oxide nanoparticles such as ZnO and TiO_2 offer enhanced photocatalytic dye degradation due to their high surface area and reactivity (Singh et al., 2018). Their small size allows better dispersion, energy efficiency, and use in remote water purification. Innovations include nano-adsorbents, catalytic membranes, and biospecific nanomaterials for removing metals, microbes, and organic pollutants (Needhidasan et al., 2017; Samuel et al., 2022).

3.3.3 Agricultural Applications

Green-synthesized nanoparticles serve as eco-friendly nanofertilizers, supplying essential nutrients and promoting plant growth. Metal NPs like Ag, Au, Cu, Zn, Fe, and Se enhance crop productivity, act as antimicrobial agents, and function as bionanosensors for disease management (Singh et al., 2015). Nanofertilizers fall into three types: (i) growth-enhancing nanomaterials like Si, TiO_2, and carbon nanotubes; (ii) micronutrient-based (Mo, Zn, Ni, Fe, Cu, Mg); and (iii) macronutrient-based (Ca, P, N, K). Phosphatic nanofertilizers have been shown to increase soybean growth and seed yield by 32% and 20%, respectively, over conventional fertilizers (Liu and Lal, 2014). AuNPs are widely used in agriculture for their antimicrobial effects and as carriers in pesticide/fertilizer delivery systems. Incorporating probiotics in the green synthesis of AgNPs and AuNPs promotes eco-friendly chemistry, avoiding harmful chemicals (Markus et al., 2016). Fungus-mediated nanoparticles, functionalized with fungal biomolecules, enhance nutrient release, stability, and absorption while reducing toxicity (Parisi et al., 2015). Such nanofertilizers also improve resistance to drought, disease, and pests, while enhancing auxin synthesis and soil nutrient solubility (Osman et al., 2024; Nanda et al., 2024).

While nanotechnology holds promise in agriculture, its use raises environmental and health concerns. To address these, global regulatory bodies have introduced safety codes and quality guidelines. The International Organization for Standardization (ISO) developed several standards, including ISO/TS 80004-1, which defines terms related to green nanotechnology (Kumari et al., 2023).

3.4 Conclusion and Future Perspectives

Green synthesis of nanoparticles is widely recognized as a safe, low-cost, and sustainable alternative to conventional methods. This chapter highlights various eco-friendly techniques for producing metal and metal oxide nanoparticles and explores their broad applications. Despite some limitations, green-synthesized NPs offer significant advantages—non-toxicity, cost-effectiveness, and enhanced biocompatibility—making them valuable in medicine, agriculture, industry, and environmental remediation by avoiding harmful synthetic chemicals.

Green-synthesized nanoparticles hold immense potential to transform sectors such as healthcare, agriculture, and environmental remediation. Researchers are actively advancing synthesis techniques to enhance nanoparticle stability, yield, and performance while minimizing their environmental impact. Vividly synthesized nanoparticles offer a promising, innovative pathway for cell-specific drug delivery, though research is still ongoing to fully explore their potential.

- Green nanoparticles enable smart fertilizers, precision pesticides, and early disease detection—boosting crop yields and minimizing waste. These innovations support global food security through sustainable practices.
- Green nanotechnologies will increasingly aid ecosystem restoration by enabling targeted pollutant removal and supporting climate mitigation efforts.
- Green nanomaterials are expected to drive sustainable industrial growth through applications in renewable energy, catalysis, and biodegradable packaging. Their potential use in fuel cells, energy storage, and eco-friendly coatings is under active investigation.
- Scaling up green nanoparticle production remains a key challenge, particularly in balancing cost and consistency. Optimizing synthesis using plant-based or waste-derived materials could enhance commercial feasibility and sustainability.

References

Abd El-Halim, M.D., Almehizia, A.A., Elasasy, M.E., Zen, A.A., Kamel, O.M. et al. 2025. Larvicidal and adulticidal effect of green synthesis copper oxide nanoparticles using *Achillea fragrantissima* and its biological and ultra-structural impact on *Culex pipiens*. *Bulletin of the Chemical Society of Ethiopia*, 39(3): 535–546.

Abishek, M.S. and Kachhap, S. 2025. Sustainable synthesis of copper oxide nanoparticles using *Spondias mombin* and biodiesel production from *Guizotia abyssinica*: Engine performance, emission characteristics, and machine learning-based optimization. *Fuel*, 392: 134804.

Abishek, M.S., Kachhap, S., Rajak, U., Verma, T.N., Giri et al. 2024. Exergy-energy, sustainability, and emissions assessment of Guizotia abyssinica (L.) fuel blends with metallic nano additives. *Scientific Reports*, 14(1): 3537.

Aboelfetoh, E.F., El-Shenody, R.A. and Ghobara, M.M. 2017. Eco-friendly synthesis of silver nanoparticles using green algae (*Caulerpa serrulata*): Reaction optimization, catalytic and antibacterial activities. *Environmental Monitoring and Assessment*, 189: 1–15.

Ahmad, T., Irfan, M. and Bhattacharjee, S. 2016. Parametric study on gold nanoparticle synthesis using aqueous *Elaise guineensis* (oil palm) leaf extract: Effect of precursor concentration. *Procedia Engineering*, 148: 1396–1401.

Ahmad, W., Jaiswal, K. K. and Soni, S. 2020. Green synthesis of titanium dioxide (TiO_2) nanoparticles by using *Mentha arvensis* leaves extract and its antimicrobial properties. *Inorganic and Nano-Metal Chemistry*, 50(10): 1032–1038.

Ahmed, S. and Ikram, S. 2016. Biosynthesis of gold nanoparticles: A green approach. *Journal of Photochemistry and Photobiology B: Biology*, 161: 141–153.

Akhtar, K., Khan, S.A., Khan, S.B., Asiri, A.M. et al. 2018. Scanning electron microscopy: Principle and applications. In *Nanomaterials Characterization* (113–145). Springer International Publishing.

Akhter, G., Khan, A., Ali, S.G., Khan, T.A., Siddiqi, K.S. et al. 2020. Antibacterial and nematicidal properties of biosynthesized Cu nanoparticles using extract of holoparasitic plant. *SN Applied Sciences*, 2: 1–6.

Alex, A.M., Subburaman, S., Chauhan, S., Ahuja, V. and Abdi, G. 2024. Green synthesis of silver nanoparticle prepared with *Ocimum* species and assessment of anticancer potential. *Scientific Reports*, 14(1): 11707.

Algotiml, R., Gab-Alla, A., Seoudi, R., Abulreesh, H.H., El-Readi, M.Z. et al. 2022. Anticancer and antimicrobial activity of biosynthesized Red Sea marine algal silver nanoparticles. *Scientific Reports*, 12(1): 2421.

Alshatwi, A.A., Athinarayanan, J. and Vaiyapuri Subbarayan, P. 2015. Green synthesis of platinum nanoparticles that induce cell death and G2/M-phase cell cycle arrest in human cervical cancer cells. *Journal of Materials Science: Materials in Medicine*, 26(7): 1–9.

Ameena, S., Rajesh, N., Anjum, S.M., Khadri, H., Riazunnisa, K. et al. 2022. Antioxidant, antibacterial, and anti-diabetic activity of green synthesized copper nanoparticles of *Cocculus hirsutus* (Menispermaceae). *Applied Biochemistry and Biotechnology*, 194(10): 4424–4438.

Amin, M.A.A., Abu-Elsaoud, A.M., Nowwar, A.I., Abdelwahab, A.T., Awad, M.A. et al. 2024. Green synthesis of magnesium oxide nanoparticles using endophytic fungal strain to improve the growth, metabolic activities, yield traits, and phenolic compounds content of *Nigella sativa* L. *Green Processing and Synthesis*, 13(1): 20230215.

Aminabad, N.S., Farshbaf, M. and Akbarzadeh, A. 2019. Recent advances of gold nanoparticles in biomedical applications: State of the art. *Cell Biochemistry and Biophysics*, 77: 123–137.

Amjad, R., Mubeen, B., Ali, S.S., Imam, S.S., Alshehri, S. et al. 2021. Green synthesis and characterization of copper nanoparticles using *Fortunella margarita* leaves. *Polymers*, 13(24): 4364.

Andreescu, S., Ornatska, M., Erlichman, J.S., Estevez, A., Leiter, J.C. et al. 2012. Biomedical applications of metal oxide nanoparticles. *Fine Particles in Medicine and Pharmacy*, 57–100.

Anwaar, S., Ijaz, D.E.S., Anwar, T., Qureshi, H., Nazish, M. et al. 2024. Boosting *Solanum tuberosum* resistance to *Alternaria solani* through green synthesized ferric oxide (Fe_2O_3) nanoparticles. *Scientific Reports*, 14(1): 2375.

Aravinth, A., Mohan, P.K., Perumal, P., Dhanasundaram, S., Kamaraj, C. et al. 2024. Phyto-fabrication of brown seaweed *Dictyota ciliolata* extract towards silver nanoparticles synthesis and evaluation of mosquito-larvicidal potential of the synthesized AgNPs. *Journal of Applied Phycology*, 36(4): 1–13.

Aromal, S.A., Vidhu, V.K. and Philip, D. 2012. Green synthesis of well-dispersed gold nanoparticles using *Macrotyloma uniflorum*. *Spectrochimica Acta Part A: Molecular and Biomolecular Spectroscopy*, 85(1): 99–104.

Asanithi, P., Chaiyakun, S. and Limsuwan, P.J.J.O.N. 2012. Growth of silver nanoparticles by DC magnetron sputtering. *Journal of Nanomaterials*, 2012(1): 963609.

Ashokkumar, M., Palanisamy, K., Ganesh Kumar, A., Muthusamy, C., Senthil Kumar, K.J. et al. 2024. Green synthesis of silver and copper nanoparticles and their composites using *Ocimum sanctum* leaf extract displayed enhanced antibacterial, antioxidant and anticancer potentials. *Artificial Cells, Nanomedicine, and Biotechnology*, 52(1): 438–448.

Azeem, M.N.A., Ahmed, O.M., Shaban, M. and Elsayed, K.N. 2022. Retracted Article: *In vitro* antioxidant, anticancer, anti-inflammatory, anti-diabetic and anti-alzheimer potentials of innovative macroalgae bio-capped silver nanoparticles. *Environmental Science and Pollution Research*, 29(39): 59930–59947.

Bai, R.G., Sabouni, R. and Husseini, G. 2018. Green nanotechnology—A road map to safer nanomaterials. In *Applications of Nanomaterials* (133–159). Woodhead Publishing. http://dx.doi.org/10.1016/B978-0-08-101971-9.00006-5.

Balaraman, P., Balasubramanian, B., Liu, W.C., Kaliannan, D., Durai, M. et al. 2022. *Sargassum myriocystum*-mediated TiO_2-nanoparticles and their antimicrobial, larvicidal activities and enhanced photocatalytic degradation of various dyes. *Environmental Research*, 204: 112278.

Baran, M.F., Keskin, C., Baran, A., Hatipoğlu, A., Yildiztekin, M. et al. 2023. Green synthesis of silver nanoparticles from *Allium cepa* L. Peel extract, their antioxidant, antipathogenic, and anticholinesterase activity. *Molecules*, 28(5): 2310.

Baskaran, X., Geo Vigila, A.V., Parimelazhagan, T., Muralidhara-Rao, D., Zhang, S. et al. 2016. Biosynthesis, characterization, and evaluation of bioactivities of leaf extract-mediated biocompatible silver nanoparticles from an early tracheophyte, *Pteris tripartita* Sw. *International Journal of Nanomedicine*, 5789–5806.

Boomi, P., Poorani, G.P., Selvam, S., Palanisamy, S., Jegatheeswaran, S. et al. 2020. Green biosynthesis of gold nanoparticles using *Croton sparsiflorus* leaves extract and evaluation of UV protection, antibacterial and anticancer applications. *Applied Organometallic Chemistry*, 34(5): e5574.

Botteon, C.E.A., Silva, L.B., Ccana-Ccapatinta, G.V., Silva, T.S., Ambrosio, S.R. et al. 2021. Biosynthesis and characterization of gold nanoparticles using Brazilian red propolis and evaluation of its antimicrobial and anticancer activities. *Scientific Reports*, 11(1): 1974.

Bruna, T., Maldonado-Bravo, F., Jara, P. and Caro, N. 2021. Silver nanoparticles and their antibacterial applications. *International Journal of Molecular Sciences*, 22(13): 7202.

Carmen, Z. and Daniela, S. 2012. Textile organic dyes-characteristics, polluting effects and separation/elimination procedures from industrial effluents—A critical overview. *Rijeka: IntechOpen*, 3: 55–86.

Castañeda-Aude, J.E., Morones-Ramírez, J.R., De Haro-Del Río, D.A., León-Buitimea, A., Barriga-Castro, E.D. et al. 2023. Ultra-small silver nanoparticles: A sustainable green synthesis approach for antibacterial activity. *Antibiotics*, 12(3): 574.

Chowdhury, N., Paramanik, M., Sarkar, N., Laskar, S., Chandra, G. et al. 2015. Fatty acid analysis of leaf and berry of *Solanum villosum* Mill. and its prospect as diet and bactericide. *Analytical Chemistry Letters*, 5(5): 260–266. https://doi.org/10.1080/22297928.2015.1137225.

Coman, N.A., Berta, L., Nicolae-Maranciuc, A., Nicolescu, A., Babotă, M. et al. 2024. Green synthesis of platinum nanoparticles using aqueous bark extract of *Quercus* sp. for potential antioxidant and antimicrobial applications. *BioResources*, 19(4): 862.

Crini, G. and Lichtfouse, E. 2018. Wastewater treatment: An overview. Green adsorbents for pollutant removal: Fundamentals and design. *Environmental Chemistry for a Sustainable World*, 18: 1–21.

Das, S.K., Das, A.R., and Guha, A.K. 2010. Microbial synthesis of multishaped gold nanostructures. *Small*, 6(9): 1012–1021.

Dasmodak, S., Paramanik, S., Mukherjee, S. and Paramanik, M. 2024a. Food chain contamination by neonicotinoids—A matter of concern. pp. 111–124. *In*: Singh, R., Singh, V.K., Kumar, A., Tripathi, S. and Bhadouria, R. (eds.). *Neonicotinoids in the Environment: Emerging Concerns for the Human Health and Biodiversity. Springer Nature, Cham, Switzerland.* https://doi.org/10.1007/978-3-031-45343-4_9.

Dasmodak, S., Paramanik, S., Kapuri, R., Mukherjee, S., Paramanik, M. et al. 2024b. A natural weapon against lymphatic filariasis and dengue vectors: The larvicidal potential of *Leonurus sibiricus* L. *International Journal of Mosquito Research*, 11(5): 92–98. https://doi.org/10.22271/23487941.2024.v11.i5b.808.

Davar, F., Majedi, A. and Mirzaei, A. 2015. Green synthesis of ZnO nanoparticles and its application in the degradation of some dyes. *Journal of the American Ceramic Society*, 98(6): 1739–1746.

Devi, T.B. and Ahmaruzzaman, M. 2018. Facile preparation of copper nanaoparticles using *Coccinia grandis* fruit extract and its application towards the reduction of toxic nitro compound. *Materials Today: Proceedings*, 5(1): 2098–2104.

Dhand, V., Soumya, L., Bharadwaj, S., Chakra, S., Bhatt, D. et al. 2016. Green synthesis of silver nanoparticles using *Coffea arabica* seed extract and its antibacterial activity. *Materials Science and Engineering: C*, 58: 36–43.

Dinga, E., Ekennia, A., Ogbonna, C.U., Udu, D.A., Mthiyane, D.M.N. et al. 2022. Phyto-mediated synthesis of MgO nanoparticles using *Melia azedarach* seed extract: Larvicidal and antioxidant activities. *Scientific African*, 17: e01366.

Doan, V.D., Thieu, A.T., Nguyen, T.D., Nguyen, V.C., Cao, X.T. et al. 2020. Biosynthesis of gold nanoparticles using *Litsea cubeba* fruit extract for catalytic reduction of 4-nitrophenol. *Journal of Nanomaterials*, 2020(1): 4548790.

Duan, L., Li, M. and Liu, H. 2015. Biosynthesised palladium nanoparticles using *Eucommia ulmoides* bark aqueous extract and their catalytic activity. *IET Nanobiotechnology*, 9(6): 349–354.

Elia, P., Zach, R., Hazan, S., Kolusheva, S., Porat, Z.E. et al. 2014. Green synthesis of gold nanoparticles using plant extracts as reducing agents. *International Journal of Nanomedicine*, 9: 4007–4021.

Fatima, R., Priya, M., Indurthi, L., Radhakrishnan, V., Sudhakaran, R. et al. 2020. Biosynthesis of silver nanoparticles using red algae *Portieria hornemannii* and its antibacterial activity against fish pathogens. *Microbial Pathogenesis*, 138: 103780.

Fouda, A., Eid, A.M., Guibal, E., Hamza, M.F., Hassan, S.E.D. et al. 2022. Green synthesis of gold nanoparticles by aqueous extract of *Zingiber officinale*: Characterization and insight into antimicrobial, antioxidant, and *in vitro* cytotoxic activities. *Applied Sciences*, 12(24): 12879.

Ghosh, N.S., Pandey, E., Giilhotra, R.M. and Singh, R. 2020. Biosynthesis of gold nanoparticles using leaf extract of *Desmodium gangeticum* and their antioxidant activity. *Research Journal of Pharmacy and Technology*, 13(6): 2685–2689.

Ghosh, S., Nitnavare, R., Dewle, A., Tomar, G.B., Chippalkatti, R. et al. 2015. Novel platinum–palladium bimetallic nanoparticles synthesized by *Dioscorea bulbifera*: Anticancer and antioxidant activities. *International Journal of Nanomedicine*, 7477–7490.

Gupta, A.K. and Gupta, M. 2005. Synthesis and surface engineering of iron oxide nanoparticles for biomedical applications. *Biomaterials*, 26: 3995–4021. https://doi.org/10.1016/j.biomaterials.2004.10.012.

Hashem, A.H., Selim, T.A., Alruhaili, M.H., Selim, S., Alkhalifah, D.H.M. et al. 2022. Unveiling antimicrobial and insecticidal activities of biosynthesized selenium nanoparticles using prickly pear peel waste. *Journal of Functional Biomaterials*, 13(3): 112.

Hiragond, C.B., Kshirsagar, A.S., Dhapte, V.V., Khanna, T., Joshi, P. et al. 2018. Enhanced antimicrobial response of commercial face mask using colloidal silver nanoparticles. *Vacuum*, 156: 475–482.

İpek, P., Baran, M.F., Baran, A., Hatipoğlu, A., Keskin, C. et al. 2024. Green synthesis and evaluation of antipathogenic, antioxidant, and anticholinesterase activities of gold nanoparticles (Au NPs) from *Allium cepa* L. peel aqueous extract. *Biomass Conversion and Biorefinery*, 14(9): 10661–10670.

Kahrilas, G.A., Wally, L.M., Fredrick, S.J., Hiskey, M., Prieto, A.L. et al. 2014. Microwave-assisted green synthesis of silver nanoparticles using orange peel extract. *ACS Sustainable Chemistry & Engineering*, 2(3): 367–376.

Khairunnisa, S., Wonoputri, V. and Samadhi, T.W. 2021. Effective deagglomeration in biosynthesized nanoparticles: A mini review. In *IOP Conference Series: Materials Science and Engineering*, 1143(1): 012006. IOP Publishing.

Klein, W., Ismail, E., Maboza, E., Hussein, A.A., Adam, R.Z. et al. 2023. Green-synthesized silver nanoparticles: Antifungal and cytotoxic potential for further dental applications. *Journal of Functional Biomaterials*, 14(7): 379.

Kumar, D., Kumar, G. and Agrawal, V. 2018. Green synthesis of silver nanoparticles using *Holarrhena antidysenterica* (L.) Wall. bark extract and their larvicidal activity against dengue and filariasis vectors. *Parasitology Research*, 117: 377–389.

Kumari, R., Suman, K., Karmakar, S., Mishra, V., Lakra, S.G. et al. 2023. Regulation and safety measures for nanotechnology-based agri-products. *Frontiers in Genome Editing*, 5: 1200987.

Latha, D., Sampurnam, S., Arulvasu, C., Prabu, P., Govindaraju, K. et al. 2018. Biosynthesis and characterization of gold nanoparticle from *Justicia adhatoda* and its catalytic activity. *Materials Today: Proceedings*, 5(2): 8968–8972.

Li, Y., Leung, P., Yao, L., Song, Q.W., Newton, E. et al. 2006. Antimicrobial effect of surgical masks coated with nanoparticles. *Journal of Hospital Infection*, 62(1): 58–63.

Liu, R. and Lal, R. 2014. Synthetic apatite nanoparticles as a phosphorus fertilizer for soybean (*Glycine max*). *Scientific Reports*, 4(1): 5686.

Loganathan, S., Govindasamy, M., Habila, M.A. and Manimaran, K. 2023. Green synthesis of iron oxide nanoparticles using *Pterolobium hexapetalum* (roth) santapau & wagh leaves extract and their biological applications. *Biocatalysis and Agricultural Biotechnology*, 54: 102905.

Madani, M., Hosny, S., Alshangiti, D.M., Nady, N., Alkhursani, S.A. et al. 2022. Green synthesis of nanoparticles for varied applications: Green renewable resources and energy-efficient synthetic routes. *Nanotechnology Reviews*, 11(1): 731–759.

Majumdar, R. and Kumar Kar, P. 2024. Green synthesis of silver nanoparticles from *Abelmoschus esculentus* (okra) calyx wastes as a promising anthelmintic agent against poultry pathogen *Raillietina* spp. *Chemistry Select*, 9(38): e202403378.

Majumdar, R., Bag, B.G. and Ghosh, P. 2016. *Mimusops elengi* bark extract mediated green synthesis of gold nanoparticles and study of its catalytic activity. *Applied Nanoscience*, 6: 521–528.

Makarov, V.V., Love, A.J., Sinitsyna, O.V., Makarova, S.S., Yaminsky, I.V. et al. 2014. "Green" nanotechnologies: Synthesis of metal nanoparticles using plants. *Acta Naturae (англоязычная версия)*, 6(1(20)): 35–44.

Managa, M., Antunes, E. and Nyokong, T. 2014. Conjugates of platinum nanoparticles with gallium tetra–(4-Carboxyphenyl) porphyrin and their use in photodynamic antimicrobial chemotherapy when in solution or embedded in electrospun fiber. *Polyhedron*, 76: 94–101.

Manjare, S.B. and Chaudhari, R.A. 2020. Palladium nanoparticle-bentonite hybrid using leaves of *Syzygium aqueum* plant from India: Design and assessment in the catalysis of–C–C–coupling reaction. *Chemistry Africa*, 3: 329–341.

Manojkumar, U., Kaliannan, D., Srinivasan, V., Balasubramanian, B., Kamyab, H. et al. 2023. Green synthesis of zinc oxide nanoparticles using *Brassica oleracea* var. botrytis leaf extract: Photocatalytic, antimicrobial and larvicidal activity. *Chemosphere*, 323: 138263.

Mansoori, G.A. 2005. *Principles of Nanotechnology: Molecular-based Study of Condensed Matter in Small Systems* (1–360). World Scientifc Publishing Company, Singapore.

Markus, J., Mathiyalagan, R., Kim, Y.J., Abbai, R., Singh, P. et al. 2016. Intracellular synthesis of gold nanoparticles with antioxidant activity by probiotic *Lactobacillus kimchicus* DCY51T isolated from Korean kimchi. *Enzyme and Microbial Technology*, 95: 85–93.

Matinise, N., Fuku, X.G., Kaviyarasu, K., Mayedwa, N., Maaza, M.J.A.S.S. et al. 2017. ZnO nanoparticles via *Moringa oleifera* green synthesis: Physical properties & mechanism of formation. *Applied Surface Science*, 406: 339–347.

McNamara, K. and Tofail, S.A. 2017. Nanoparticles in biomedical applications. *Advances in Physics: X*, 2(1): 54–88.

Medina-Acosta, M., Chinchillas-Chinchillas, M.J., Garrafa-Gálvez, H.E., Garcia-Maro, C.A., Rosas-Casarez, C.A. et al. 2024. Photocatalytic degradation of four organic dyes present in water using ZnO nanoparticles synthesized with green synthesis using *Ambrosia ambrosioides* leaf and root extract. *Processes*, 12(11): 2456.

Mohanta, Y.K., Biswas, K., Jena, S.K., Hashem, A., Abd_Allah, E.F. et al. 2020. Anti-biofilm and antibacterial activities of silver nanoparticles synthesized by the reducing activity of phytoconstituents present in the Indian medicinal plants. *Frontiers in Microbiology*, 11: 1143.

Muhaymin, A., Mohamed, H.E.A., Hkiri, K., Safdar, A., Azizi, S. and Maaza, M. 2024. Green synthesis of magnesium oxide nanoparticles using *Hyphaene thebaica* extract and their photocatalytic activities. *Scientific Reports*, 14(1): 20135.

Murthy, H.A., Desalegn, T., Kassa, M., Abebe, B., Assefa, T. et al. 2020. Synthesis of green copper nanoparticles using medicinal plant *Hagenia abyssinica* (Brace) JF. Gmel. leaf extract: Antimicrobial properties. *Journal of Nanomaterials*, 2020(1): 3924081.

Nakhjavani, M., Nikkhah, V., Sarafraz, M.M., Shoja, S., Sarafraz, M. et al. 2017. Green synthesis of silver nanoparticles using green tea leaves: Experimental study on the morphological, rheological and antibacterial behaviour. *Heat and Mass Transfer*, 53: 3201–3209.

Nanda, S., Ganguly, A., Sarkar, S., Mandi, M., Das, K. et al. 2024. Nanopesticides in agriculture: Scopes & limitations. *In*: Mabrouki, J. and Mourade, A. (eds.). *Technical and Technological Solutions Towards a Sustainable Society and Circular Economy, World Sustainability Series. Springer, Cham.* pp. 179–193. https://doi.org/10.1007/978-3-031-56292-1_14.

Narayanan, K.B. and Sakthivel, N. 2011. Green synthesis of biogenic metal nanoparticles by terrestrial and aquatic phototrophic and heterotrophic eukaryotes and biocompatible agents. *Advances in Colloid and Interface Science*, 169(2): 59–79.

Needhidasan, S. and Ramalingam, C. 2017. Stratagems employed for 2, 4-dichlorophenoxyacetic acid removal from polluted water sources. *Clean Technologies and Environmental Policy*, 19: 1607–1620.

Osman, A.I., Zhang, Y., Farghali, M., Rashwan, A.K., Eltaweil, A.S. et al. 2024. Synthesis of green nanoparticles for energy, biomedical, environmental, agricultural, and food applications: A review. *Environmental Chemistry Letters*, 22(2): 841–887.

Paramanik, M. 2018. Bio efficiency of some medicinal plant's leaf powder against the stored grain pest *Sitophilus oryzae* L. *Wesleyan Journal of Research*, 11(1): 4–8.

Paramanik, M., Chatterjee, S.K. and Chandra, G. 2023b. Larvicidal efficacy of the *Vangueria spinosa* Roxb. (Rubiaceae) leaf extracts against filarial vector *Culex quinquefasciatus. International Journal of Mosquito Research*, 10(1): 01–06. https://doi.org/10.22271/23487941.2023.v10.i1a.655.

Paramanik, M., Paramanik, S., Mukherjee, S., Kapuri, R., Dasmodak, S. et al. 2022. Larvicidal efficacy of the leaf and flower extracts of *Tropaeolum majus* L. (Tropaeolaceae) against the vector of lymphatic filariasis. *International Journal of Mosquito Research*, 9(2): 26–29. https://doi.org/10.22271/23487941.2022.v9.i2a.597.

Paramanik, S., Dasmodak, S., Mukherjee, S., Rajak, P., Paramanik, M. et al. 2024a. Non-target toxicity of neonicotinoid insecticides impeding the behaviour and population of bees. pp. 63–73. *In*: Singh, R., Singh, V.K., Kumar, A., Tripathi, S. and Bhadouria, R. (eds.). *Neonicotinoids in the Environment: Emerging Concerns for the Human Health and Biodiversity. Springer Nature, Cham, Switzerland.* https://doi.org/10.1007/978-3-031-45343-4_5.

Paramanik, S., Ganguly, A., Jana, K., Rajak, P., Paramanik, M. et al. 2023a. Nanotechnology: A novel weapon for insect pest and vector management. pp. 277–296. *In*: Verma, R.K., Singh, D.K. and Davim, J.P. (eds.). *Nanocomposite and Nanohybrid Materials: Processing and Applications. Walter De Gruyter, Berlin*, 17. https://doi.org/10.1515/9783111137902-014.

Paramanik, S., Mukherjee, S. and Paramanik, M. 2024. Phytoconstituent analysis of petroleum ether leaf extract of *Olax scandens* and study of its mosquito larvicidal potency. *Science and Culture*, 90(7-8): 306–309. https://doi.org/10.36094/sc.v89.2024.Phytoconstituent_analysis_of_Petroleum.Paramanik.306.

Parisi, C., Vigani, M. and Rodríguez-Cerezo, E. 2015. Agricultural nanotechnologies: What are the current possibilities? *Nano Today*, 10(2): 124–127.

Patra, J.K. and Baek, K.H. 2014. Green nanobiotechnology: Factors affecting synthesis and characterization techniques. *Journal of Nanomaterials*, 2014(1): 417305.

Patra, N., Dehury, N., Pal, A., Behera, A., Patra, S. et al. 2018. Preparation and mechanistic aspect of natural xanthone functionalized gold nanoparticle. *Materials Science and Engineering: C*, 90: 439–445.

Paul, K., Bag, B.G. and Samanta, K. 2014. Green coconut (*Cocos nucifera* Linn) shell extract mediated size controlled green synthesis of polyshaped gold nanoparticles and its application in catalysis. *Applied Nanoscience*, 4: 769–775.

Philip, D. 2009. Honey mediated green synthesis of gold nanoparticles. *Spectrochimica Acta Part A: Molecular and Biomolecular Spectroscopy*, 73(4): 650–653.

Pirtarighat, S., Ghannadnia, M. and Baghshahi, S. 2019. Green synthesis of silver nanoparticles using the plant extract of *Salvia spinosa* grown *in vitro* and their antibacterial activity assessment. *Journal of Nanostructure in Chemistry*, 9: 1–9.

Ponarulselvam, S., Panneerselvam, C., Murugan, K., Aarthi, N., Kalimuthu, K. et al. 2012. Synthesis of silver nanoparticles using leaves *of Catharanthus roseus* Linn. G. Don and their antiplasmodial activities. *Asian Pacific Journal of Tropical Biomedicine*, 2(7): 574–580.

Prakash, P., Gnanaprakasam, P., Emmanuel, R., Arokiyaraj, S., Saravanan, M. et al. 2013. Green synthesis of silver nanoparticles from leaf extract of *Mimusops elengi*, Linn. for enhanced antibacterial activity against multi drug resistant clinical isolates. *Colloids and Surfaces B: Biointerfaces*, 108: 255–259.

Rajesh kumar, S., Malarkodi, C., Vanaja, M., Gnanajobitha, G., Paulkumar et al. 2013. Antibacterial activity of algae mediated synthesis of gold nanoparticles from *Turbinaria conoides*. *Der Pharma Chemica*, 5(2): 224–229.

Rajesh Kumar, T.V., Murthy, J.S.R., Narayana Rao, M. and Bhargava, Y. 2016. Evaluation of silver nanoparticles synthetic potential of *Couroupita guianensis* Aubl., flower buds extract and their synergistic antibacterial activity. *3 Biotech.*, 6(1): 92.

Rajkumar, M., Presley, S.D., Menaa, F., Elbehairi, S.E.I., Alfaifi, M.Y. et al. 2024. Biosynthesis and biological activities of magnesium hydroxide nanoparticles using *Tinospora cordifolia* leaf extract. *Bioprocess and Biosystems Engineering*, 1–19.

Rastogi, L., Beedu, S.R. and Kora, A.J. 2015. Facile synthesis of palladium nanocatalyst using gum kondagogu (*Cochlospermum gossypium*): A natural biopolymer. *IET Nanobiotechnology*, 9(6): 362–367.

Rautela, A. and Rani, J. 2019. Green synthesis of silver nanoparticles from *Tectona grandis* seeds extract: Characterization and mechanism of antimicrobial action on different microorganisms. *Journal of Analytical Science and Technology*, 10(1): 1–10.

Raveendran, P., Fu, J. and Wallen, S.L. 2003. Completely "green" synthesis and stabilization of metal nanoparticles. *Journal of the American Chemical Society*, 125(46): 13940–13941.

Roe, D., Karandikar, B., Bonn-Savage, N., Gibbins, B., and Roullet, J.B. 2008. Antimicrobial surface functionalization of plastic catheters by silver nanoparticles. *Journal of Antimicrobial Chemotherapy*, 61(4): 69–876.

Samiei, A., Tavassoli, M. and Esmaeilnejad, B. 2024. Green synthesis and anthelmintic activity of silver nanoparticles using *Morus alba* Fruit extract against different stages of *Equine strongyles*. *Veterinary Research Communications*, 48(4): 1–16.

Samuel, M.S., Ravikumar, M., John, J.A., Selvarajan, E., Patel, H. et al. 2022. A review on green synthesis of nanoparticles and their diverse biomedical and environmental applications. *Catalysts*, 12(5): 459.

Sana, S.S. and Dogiparthi, L.K. 2018. Green synthesis of silver nanoparticles using *Givotia moluccana* leaf extract and evaluation of their antimicrobial activity. *Materials Letters*, 226: 47–51.

Sangameshwaran, V., Prasath, A.R., Selvam, K., Obaidullah, A.J., Yadav, K.K. et al. 2024. Green synthesis of silver nanoparticles using *Senna alexandrina* leaf extracts and their antibacterial, larvicidal, nanotoxicity and photocatalytic activity. *Inorganica Chimica Acta*, 571: 122220.

Sathishkumar, G., Jha, P.K., Vignesh, V., Rajkuberan, C., Jeyaraj, M. et al. 2016. Cannonball fruit (*Couroupita guianensis*, Aubl.) extract mediated synthesis of gold nanoparticles and evaluation of its antioxidant activity. *Journal of Molecular Liquids*, 215: 229–236.

Shah, M.Z., Guan, Z.H., Din, A.U., Ali, A., Rehman, A.U. et al. 2021. Synthesis of silver nanoparticles using *Plantago lanceolata* extract and assessing their antibacterial and antioxidant activities. *Scientific Reports*, 11(1): 20754.

Sharma, B.K., Shah, D.V. and Roy, D.R. 2018. Green synthesis of CuO nanoparticles using *Azadirachta indica* and its antibacterial activity for medicinal applications. *Materials Research Express*, 5(9): 095033.

Sharma, N.K., Vishwakarma, J., Rai, S., Alomar, T.S., AlMasoud, N. et al. 2022. Green route synthesis and characterization techniques of silver nanoparticles and their biological adeptness. *ACS Omega*, 7(31): 27004–20.

Sharmila, G., Haries, S., Fathima, M.F., Geetha, S., Kumar, N.M. et al. 2017. Enhanced catalytic and antibacterial activities of phytosynthesized palladium nanoparticles using *Santalum album* leaf extract. *Powder Technology*, 320: 22–26.

Shiny, P.J., Mukherjee, A. and Chandrasekaran, N. 2014. Haemocompatibility assessment of synthesised platinum nanoparticles and its implication in biology. *Bioprocess and Biosystems Engineering*, 37: 991–997.

Shivaramakrishnan, B., Gurumurthy, B. and Balasubramanian, A. 2017. Potential biomedical applications of metallic nanobiomaterials: A review. *International Journal of Pharmaceutical Sciences and Research*, 8(3): 985.

Singh, A.K. and Srivastava, O. 2015. One-step green synthesis of gold nanoparticles using black cardamom and effect of pH on its synthesis. *Nanoscale Research Letters*, 10: 1–12.

Singh, H., Desimone, M.F., Pandya, S., Jasani, S., George, N. et al. 2023. Revisiting the green synthesis of nanoparticles: Uncovering influences of plant extracts as reducing agents for enhanced synthesis efficiency and its biomedical applications. *International Journal of Nanomedicine*, 18: 4727–4750.

Singh, J., Dutta, T., Kim, K.H., Rawat, M., Samddar, P. et al. 2018. 'Green'synthesis of metals and their oxide nanoparticles: Applications for environmental remediation. *Journal of Nanobiotechnology*, 16: 1–24.

Singh, S., Singh, B.K., Yadav, S.M. and Gupta, A.K. 2015. Applications of nanotechnology in agricultural and their role in disease management. *Research Journal of Nanoscience and Nanotechnology*, 5(1): 1–5.

Singh, V., Koshti, B. and Gour, N. 2021. Green synthesized nanoparticles: Classification, synthesis, characterization, and applications. *Comprehensive Analytical Chemistry*, 94: 173–222. http://dx.doi.org/10.1016/bs.coac.2020.12.009.

Sinha, T. and Ahmaruzzaman, M. 2015. Green synthesis of copper nanoparticles for the efficient removal (degradation) of dye from aqueous phase. *Environmental Science and Pollution Research*, 22: 20092–20100.

Smitha, S.L., Philip, D. and Gopchandran, K.G. 2009. Green synthesis of gold nanoparticles using *Cinnamomum zeylanicum* leaf broth. *Spectrochimica Acta Part A: Molecular and Biomolecular Spectroscopy*, 74(3): 735–739.

Sneddon, G.C., Trimby, P.W. and Cairney, J.M. 2016. Transmission Kikuchi diffraction in a scanning electron microscope: A review. *Materials Science and Engineering: R: Reports*, 110: 1–12.

Soliman, M.K., Abu-Elghait, M., Salem, S.S. and Azab, M.S. 2024. Multifunctional properties of silver and gold nanoparticles synthesis by *Fusarium pseudonygamai*. *Biomass Conversion and Biorefinery*, 14(22): 28253–28270.

Sotiriou, G.A. and Pratsinis, S.E. 2010. Antibacterial activity of nanosilver ions and particles. *Environmental Science & Technology*, 44(14): 5649–5654.

Srikar, S.K., Giri, D.D., Pal, D.B., Mishra, P.K., Upadhyay, S.N. et al. 2016. Green synthesis of silver nanoparticles: A review. *Green and Sustainable Chemistry*, 6(1): 34.

Suárez-Cerda, J., Espinoza-Gómez, H., Alonso-Núñez, G., Rivero, I.A., Gochi-Ponce, Y. et al. 2017. A green synthesis of copper nanoparticles using *Native cyclodextrins* as stabilizing agents. *Journal of Saudi Chemical Society*, 21(3): 341–348.

Sunderam, V., Thiyagarajan, D., Lawrence, A.V., Mohammed, S.S.S., Selvaraj, A. et al. 2019. *In-vitro* antimicrobial and anticancer properties of green synthesized gold nanoparticles using *Anacardium occidentale* leaves extract. *Saudi Journal of Biological Sciences*, 26(3): 455–459.

U Din, M.M., Batool, A., Ashraf, R.S., Yaqub, A., Rashid, A. et al. 2024. Green synthesis and characterization of biologically synthesized and antibiotic-conjugated silver nanoparticles followed by post-synthesis assessment for antibacterial and antioxidant applications. *ACS Omega*, 9(17): 18909–18921.

Ulbrich, K., Hola, K., Subr, V., Bakandritsos, A., Tucek, J. et al. 2016. Targeted drug delivery with polymers and magnetic nanoparticles: Covalent and noncovalent approaches, release control, and clinical studies. *Chemical Reviews*, 116(9): 5338–5431.

Velsankar, K., Sudhahar, S. and Maheshwaran, G. 2019. Effect of biosynthesis of ZnO nanoparticles via *Cucurbita* seed extract on *Culex tritaeniorhynchus* mosquito larvae with its biological applications. *Journal of Photochemistry and Photobiology B: Biology*, 200: 111650.

Vijayaram, S., Razafindralambo, H., Sun, Y.Z., Vasantharaj, S., Ghafarifarsani, H. et al. 2024. Applications of green synthesized metal nanoparticles—A review. *Biological Trace Element Research*, 202(1): 360–386.

Vinothkanna, A., Mathivanan, K., Ananth, S., Ma, Y., Sekar, S. et al. 2023. Biosynthesis of copper oxide nanoparticles using *Rubia cordifolia* bark extract: characterization, antibacterial, antioxidant, larvicidal and photocatalytic activities. *Environmental Science and Pollution Research*, 30(15): 42563–42574.

Wang, D., Xue, B., Wang, L., Zhang, Y., Liu, L. et al. 2021. Fungus-mediated green synthesis of nano-silver using *Aspergillus sydowii* and its antifungal/antiproliferative activities. *Scientific Reports*, 11(1): 10356.

Wang, M., Meng, Y., Zhu, H., Hu, Y., Xu, C.P. et al. 2021. Green synthesized gold nanoparticles using *Viola betonicifolia* leaves extract: Characterization, antimicrobial, antioxidant, and cytobiocompatible activities. *International Journal of Nanomedicine*, 16: 7319–7337.

Wani, A.K., Akhtar, N., Mir, T.U.G., Singh, R., Jha, P.K. et al. 2023. Targeting apoptotic pathway of cancer cells with phytochemicals and plant-based nanomaterials. *Biomolecules*, 13(2): 194.

Wei, L., Lu, J., Xu, H., Patel, A., Chen, Z.S. et al. 2015. Silver nanoparticles: Synthesis, properties, and therapeutic applications. *Drug Discovery Today*, 20(5): 595–601.

Wilson, J.J., Mahendran, S., Sivakumar, T., Ponmanickam, P., Thangaraj, R. et al. 2023. Mosquito larvicidal activity of silver nanoparticles synthesized using *Azolla pinnata* against *Culex quinquefasciatus* say (Diptera: Culicidae). *South African Journal of Botany*, 157: 380–386.

Yayintas, O.T., Demir, N., Canbolat, F., Ayna, T.K., Pehlivan, M. et al. 2024. Characterization, biological activity, and anticancer effect of green-synthesized gold nanoparticles using *Nasturtium officinale* L. *BMC Complementary Medicine and Therapies*, 24(1): 346.

Younis, I.Y., El-Hawary, S.S., Eldahshan, O.A., Abdel-Aziz, M.M., Ali, Z.Y. et al. 2021. Green synthesis of magnesium nanoparticles mediated from *Rosa floribunda charisma* extract and its antioxidant, antiaging and antibiofilm activities. *Scientific Reports*, 11(1): 16868.

CHAPTER

4 Innovations in Biomaterials

Neslihan Tamsü Selli,[1] *Alpay Tamer Ertürk*[2,*] and *Neslihan Başaran*[1]

This chapter offers a thorough exploration of the evolving field of biomaterials, emphasizing their transformative impact on modern healthcare and regenerative medicine. It begins with an overview of biomaterials, outlining their historical development and classifying them into polymers, ceramics, metals, and composites. The narrative then shifts to recent advances, particularly in polymer-based biomaterials, with a focus on biodegradable and biocompatible polymers vital for applications like soft tissue repair. Smart polymers—responsive to environmental stimuli such as temperature and pH—are highlighted for their potential in personalized therapies. The chapter also examines ceramic and composite materials used in orthopedic and dental contexts, underscoring their mechanical strength and bioactivity. Developments in metallic biomaterials are discussed, including new alloy formulations and surface modification techniques to improve corrosion resistance and biomechanical integration. Additionally, the chapter explores emerging fabrication methods, such as additive manufacturing and bioprinting, which enable patient-specific implants and functional bioinks for tissue engineering. It concludes by addressing future trends, ethical challenges, and regulatory frameworks crucial for translating these innovations into clinical practice.

[1] Department of Materials Science and Engineering, Gebze Technical University, Turkey.
[2] Mechanical Engineering Department, Kocaeli University, Umuttepe Campus, Turkey.
* Corresponding author: tamer.erturk@kocaeli.edu.tr
ORCID IDs of authors: https://orcid.org/0000-0002-1871-2925; https://orcid.org/0000-0002-2901-5703; https://orcid.org/0000-0001-7467-861

DOI: 10.1201/9781003607359-4

4.1 Introduction and Fundamental Concepts

4.1.1 Definition and History of Biomaterials

Biomaterials—whether natural or engineered—are designed to interact with biological systems for medical purposes. Their applications range from implants and prosthetics to drug delivery systems and tissue engineering. Selection depends primarily on biocompatibility and on mechanical, chemical, and physical properties suited to specific medical needs (Li et al., 2020). The evolution of biomaterials closely mirrors advancements in medicine and technology, with innovations significantly improving treatment efficacy and demonstrating their adaptability—from basic implants to cutting-edge tissue regeneration techniques (Schneider et al., 2023). Historically, resilient materials have long played a role in medical care; for instance, the Romans, Chinese, and Aztecs employed gold in dental procedures (Marin, 2023). Historically, materials such as ivory, wood, and glass were ingeniously used in early medical implants, including for repairing bone fractures. The widespread use of ivory and metals in artificial body parts underscores their durability and functional value. By the early 18th century, dentists were already utilizing metals like gold for dental crowns. The 19th century marked a turning point with the introduction of silver and aluminium in medical applications. However, it was the parallel advancement of sterilization techniques that truly revolutionized metal use in surgery, enhancing both safety and reliability. The modern era of biomaterials began in the 20th century with the emergence of synthetic polymers, ceramics, and composites—materials that offered superior flexibility, strength, and biocompatibility. The introduction of stainless steel and titanium implants transformed orthopedic surgery, enabling the development of more durable and functional artificial joints. Figure 4.1 presents a chronological timeline of biomaterial development, while Fig. 4.2 traces the evolution of these materials

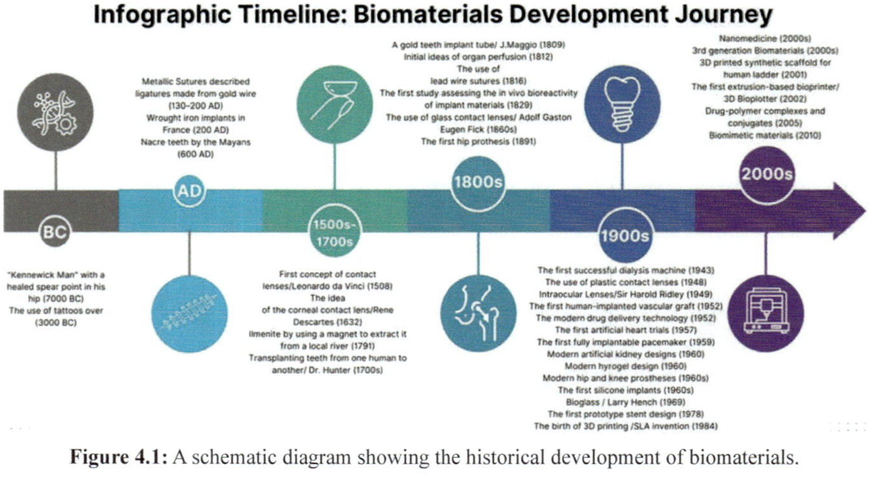

Figure 4.1: A schematic diagram showing the historical development of biomaterials.

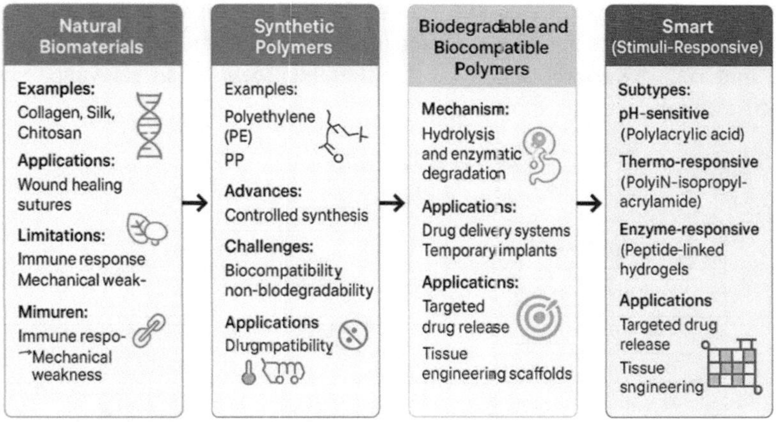

Figure 4.2: The evolution of biomaterials from natural origins to smart polymers.

from natural origins to smart polymers, highlighting major milestones, mechanisms, applications, and associated challenges.

4.1.2 Current Biomaterial Categories

Biomaterials are broadly categorized into polymers, ceramics, metals, and composites—each offering unique advantages for specific medical applications. Natural and synthetic polymers are widely used in soft tissue repair and drug delivery due to their adaptability and biocompatibility. Ceramics, prized for their strength and bioactivity, show strong potential in bone and dental regeneration. Metals like titanium and stainless steel are favored in implants for their durability, structural integrity, and corrosion resistance. Composite materials, which combine different components to improve mechanical and biological properties, are increasingly applied in orthopedics and tissue engineering. Ongoing innovations—such as nanoscale materials and bioactive coatings—continue to enhance the functionality and therapeutic value of modern biomaterials (Farag, 2023).

4.2 Advanced Polymer-Based Biomaterials

4.2.1 Biodegradable and Biocompatible Polymers

Biodegradable and biocompatible polymers represent a pivotal area in materials science, offering practical solutions for biomedical engineering and sustainable packaging. Designed to decompose into non-toxic byproducts within biological or environmental systems, they address the growing demand for eco-friendly alternatives. Their compatibility with living tissues makes them essential for drug delivery, tissue regeneration, and implantable medical devices. By combining functionality with environmental responsibility, these polymers play a vital role in advancing green technologies (Satchanska et al., 2024). Figure 4.2 classifies these materials by source and type, highlighting their broad relevance across scientific and industrial sectors.

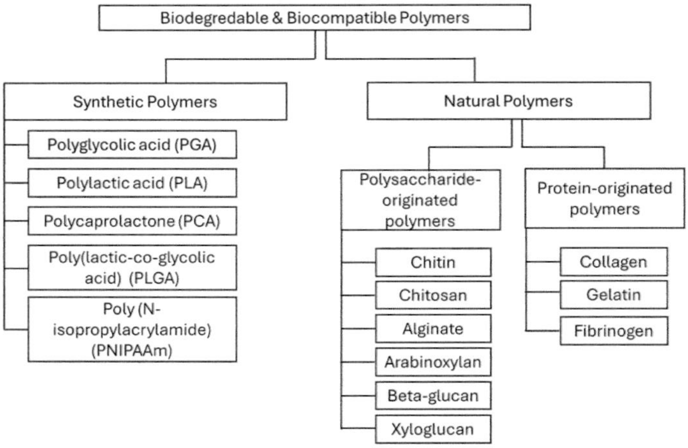

Figure 4.3: Classification of biodegradable and biocompatible polymers.

Studies related to this topic in the literature are summarized in Table 4.1. These polymers can be utilized individually or blended with other polymeric materials, offering advantages in processing and shaping for various applications.

4.2.2 Smart and Adaptive Polymers

Smart and adaptive polymers are innovative materials that respond dynamically to external stimuli such as temperature, pH, light, and mechanical forces. Their ability to reversibly modify their physical, chemical, or mechanical properties makes them valuable across diverse domains, including biomedicine, electronics, and advanced manufacturing. Mimicking biological systems, these materials enable cutting-edge developments like self-healing materials, targeted drug delivery systems, and shape-memory technologies. As shown in Fig. 4.3, smart and adaptive polymers are broadly categorized into two types: pH-sensitive and temperature-sensitive biopolymers.

4.2.2.1 pH-Sensitive Biopolymers

In pH-sensitive polymers, shifts in pH alter the ionization of functional groups, enabling researchers to control hydrogel behavior and regulate the uptake and release of particles or biomolecules. This responsiveness is governed by the ionization of pendant groups, which undergo abrupt changes near their pKa, leading to rearrangements in the polymer network. These changes produce macroscopic effects such as micelle or gel formation, swelling variations, and

Table 4.1: Selected studies on biodegradable and biocompatible polymers.

Biopolymers composition	**Properties**	**Production technique**	**Application area**	**References**
Gelatin-Polylactic Acid (PLA)	Reduces inflammation; enhances cell proliferation and attachment	Freeze drying	Cartilage repair	Fattahi et al., 2019 Fattahi et al., 2020
PC-poly(L-lactide-coglycolide)-triphosphate	Promotes new bone formation and osseointegration	AM bio-printing	Bone construction	Erten et al., 2018
Collagen- Poly(L-lactic acid) (PLLA)	Facilitates *in-vitro* bone formation	Electrospinning	Bone regeneration	Sha et al., 2016
Poly-hydroxyapatite (PHA)/PLA	Addresses inflammation; supports bone repair	AM bio-printing	Bone repair	Jaidev et al., 2019
PLGA/HA (Hydroxy-apatite)	Stimulates osteoblast activity	Biotemplating	Bone tissue engineering	Qian et al., 2014
Chitosan	Promotes cell proliferation due to high bioactivity and cytocompatibility	Freeze-drying	Bone tissue engineering	Jindal et al., 2020
Gelatin	Biocompatible and bioactive; interacts well with biological systems	Freeze-drying	Tissue Engineering	Olad et al, 2019
Gelatin & Polycaprolactone (PCL)	Shows strong biocompatibility with chondrocytes	AM bio-printing	Cartilage regeneration	Zheng et al., 2014
Alginate	Low immunogenicity; Supports cell-friendly environments	AM bio-printing	Tissue engineering, wound healing	Tomić et al., 2023
Collagen	Enhances cell attachment and growth	AM bio-printing	Tissue engineering, wound healing	Dong et al., 2016

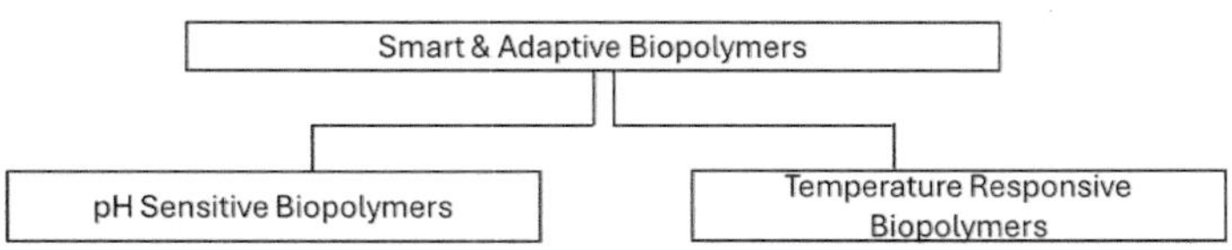

Figure 4.4: Classification of smart and adaptive biopolymers.

sol–gel transitions. Given the body's pH range—from highly acidic in the stomach (pH 1–2) to near-neutral in the intestines (pH 6–7.4)—pH-responsive biopolymers are ideal for targeted biomedical applications. Their smart mechanisms typically include polymer swelling, sol–gel phase shifts, and hydrophilic–hydrophobic transitions (Altomare et al., 2018).

4.2.2.2 Temperature Responsive Biopolymers

Temperature-responsive polymers, which undergo significant macroscopic changes in response to thermal stimuli, have emerged as powerful tools for advanced functional systems (Melone et al., 2012). Their effectiveness depends largely on alignment between their phase transition temperature and the target application's thermal conditions (De Nardo et al., 2010). Thanks to their tunable molecular structures and modifiable phase behavior, these biopolymers are receiving growing research attention (De Nardo et al., 2011). Among them, temperature-sensitive hydrogels stand out for their thermally driven gelation, swelling, and changes in hydrophilicity. These hydrogels are generally divided into two categories: (1) upper critical solution temperature (UCST) types, which swell more in compatible solvents at higher temperatures due to abundant hydrophilic groups, and (2) lower critical solution temperature (LCST) types.

Temperature-responsive biopolymers exhibit smart behavior through mechanisms such as sol-gel transitions and hydrophilic–hydrophobic phase changes. Table 4.2 presents selected studies on pH-sensitive, and temperature responsive biopolymers.

Table 4.2: This table summarizes smart and adaptive biopolymers categorized by their composition, response mechanism, application area, and supporting references.

Biopolymer type	**Composition**	**Response mechanism**	**Application area**	**References**
Temperature Responsive	Methylcellulose (MC)	Sol–gel transition	Nerve gap injuries	Wells et al., 1997
Temperature Responsive	MC + Hydroxyapatite (HA)	Sol–gel transition	Spinal cord injury treatment	Gupta et al., 2006; Kang et al., 2009
Temperature Responsive	Alginate + MC	Sol–gel transition	Bone tissue engineering	Gialouri et al., 2023
Temperature Responsive	Methylcellulose (MC) + Hydroxyapatite (HA)	Sol–gel transition	Retinal degenerative disease treatment	Ballios et al., 2010
Temperature Responsive	PCL-pluronic-chitosan-gelatin	Sol–gel transition	AM-printed porous scaffolds	Koumentakou et al., 2024
Temperature Responsive	Methylcellulose (MC)	Hydrophilic/ hydrophobic transition	Cell sheet production	Contessi et al., 2017; Chen et al., 2006
Temperature Responsive	Xyloglucan	Hydrophilic/ hydrophobic transition	Cell sheet production	Silva et al., 2013
pH-Sensitive	Chitosan	Sol–gel transition	Injectable biomaterial	Chiu et al., 2009
pH-Sensitive	Alginate	Sol–gel transition	Drug delivery	Guo et al., 2016
pH-Sensitive	Melanin	Polymer relaxation	Drug delivery	Araujo et al., 2014
pH-Sensitive	Alginate + Gelatin	Polymer relaxation	Drug delivery	Dong et al., 2006
pH-Sensitive	Pectin	Swelling	Biomaterial	Burke et al., 2003
pH-Sensitive	Alginate + Chitosan	Swelling	Drug delivery	Maity et al., 2017
pH-Sensitive	Collagen	Swelling	Drug delivery	Saito et al., 2007

4.3 Ceramic and Composite Biomaterials

Ceramic and composite biomaterials play a vital role in biomedical applications due to their excellent mechanical strength, biocompatibility, and multifunctional properties. Hydroxyapatite (HA) and bioactive glasses are widely used in bone regeneration and dental prosthetics for their structural similarity to natural bone. Composite biomaterials—formed by combining ceramics with polymers or metals—are engineered to deliver enhanced toughness, flexibility, and biological integration tailored to specific clinical needs.

4.3.1 Ceramics and Composites for Bone and Dental Regeneration

Ceramic materials such as hydroxyapatite and bioactive glasses closely mimic the chemical composition and structure of natural bone mineral, promoting effective osseointegration and new bone formation. Composite systems—formed by combining ceramics with polymeric or metallic matrices—offer enhanced mechanical strength and tunable biodegradation, making them ideal for long-term implants and tissue-engineered scaffolds (Bavya et al., 2019; Tanvir et al., 2024). Figure 4.4 schematically illustrates the role of these materials in bone and dental tissue regeneration.

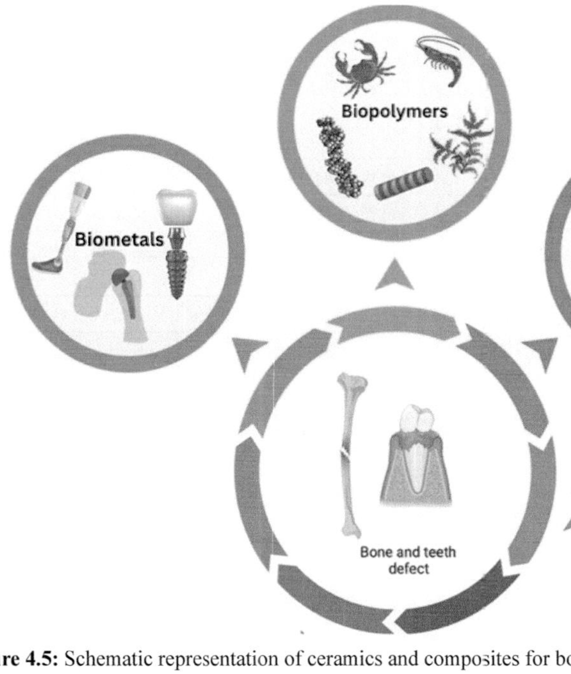

Figure 4.5: Schematic representation of ceramics and composites for bone and dental regeneration.

Bioceramic materials are typically classified into three main categories based on their interaction with biological environments: inert, bioactive, and resorbable ceramics. Inert ceramics—such as alumina and zirconia—exhibit minimal biological reactivity and are widely used in load-bearing orthopedic implants due to their high mechanical strength and wear resistance. In contrast, bioactive ceramics like hydroxyapatite and bioactive glasses form direct chemical bonds with bone, promoting osseointegration and hard tissue regeneration. Resorbable ceramics, such as tricalcium phosphate, gradually degrade in physiological conditions and are progressively replaced by native tissue. This classification enables the tailored selection of bioceramics for specific clinical needs, from bone grafts to dental implants. Despite their strengths in biocompatibility and mechanical performance, bioceramics suffer from brittleness and limited fracture toughness, constraining their use in dynamic, load-bearing settings. To address these limitations, biocomposites—combinations of ceramics with polymers, metals, or other materials—have been developed to enhance toughness and functional versatility. Table 4.3 outlines key studies on bioceramic and composite materials in bone regeneration.

Table 4.3: Selected studies on ceramics and composite materials for bone regeneration.

Material type	Composition	Application area	References
Bioceramics	Alumina	Hip replacement	Hennes et al., 2004
	Titanium nitride	Knee prostheses	Heimann et al., 2013
	Silicon nitride	Femoral heads	Olofsson et al., 2012; Mazzocchi et al., 2008
	Zirconium nitride	Knee prostheses	Heimann et al., 2013
	Bioglass	Bone replacement	Hench et al., 1991
	Hydroxyapatite (HA)	Bone cavity fillers	Cao et al, 1996; Lobel et al., 1998
Biocomposites	HA + PCL	Tissue engineering scaffolds	Hutmacher et al., 2007
	$Zn_3(PO_4)_2 + Ti(HPO_4)_x \cdot yH_2O$	Osteoporotic bone regeneration	Shrestha et al., 2022
	β-TCP + PGA	Porous composite scaffold	Cao et al., 2010
	PLGA + HA	Porous composite scaffold	Huang et al., 2008
	HA+ Ag particles	Bone regeneration	Zhang et al., 1997
	$ZrO_2 + Al_2O_3$	Joint prosthesis	De Aza et al., 2002
	ZrO_2 + HA	Orthopedic applications	Zhang et al., 2006
	SiC+HA	Load-bearing orthopedic implants	Zhao et al., 2008
	Carbon Fiber (CF) + HA	Implant materials	Zhang et al, 2003
	HA + (Strontium) Sr particles	Bone cement	Guo et al., 2011
	HA+ Sr + Ca particles		Guo et al., 2005

Teeth are composed of distinct anatomical sections—including enamel, dentin, pulp cavity, gums, root canal, bone, cementum, nerves, and blood vessels (Fig. 4.5). Each of these regions necessitates specific bioceramic and composite materials tailored to support treatment and the regeneration of damaged tissues.

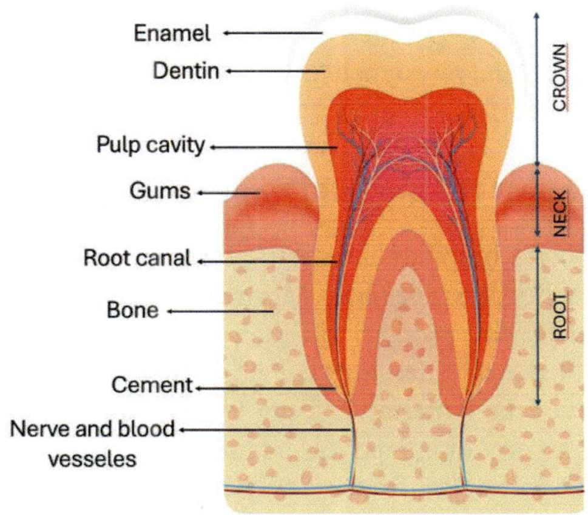

Figure 4.6: Anatomy and parts of the tooth.

Each dental structure requires biomaterials specifically designed to meet its unique biological and mechanical demands. These materials help restore function, support tissue regeneration, and promote the long-term health of teeth and surrounding tissues (Table 4.4).

Table 4.4: Biomaterials used in tooth structures and their functions.

Tooth part	Biomaterials used	Purpose
Enamel	Resin composites, ceramics, bioactive glass	Restore enamel, repair cavities, enhance aesthetics
Dentin	Hydroxyapatite, glass ionomers	Support remineralization, structural integrity, and dentin sealing
Pulp Cavity	Biocompatible scaffolds, growth factors, stem cells	Regenerate pulp tissue and restore vitality
Gums (Gingiva)	Collagen membranes, guided tissue regeneration (GTR) materials	Promote soft tissue healing and periodontal regeneration
Root Canal	Gutta-percha, MTA, bioceramic sealers	Seal canal, prevent reinfection, and maintain long-term stability
Bone (Alveolar)	Hydroxyapatite, β-TCP, bioactive glass	Support bone regeneration and implant stabilization
Cementum	Bioactive glass, calcium phosphate-based cements	Repair root surface and tissue attachment
Nerve	Biomimetic scaffolds, neuroactive polymers	Facilitate nerve regeneration (under experimental investigation)
Blood Vessels	Angiogenic growth factors, vascular scaffolds	Stimulate vascularization for enhanced healing and integration

Bioactive ceramics are widely employed in dental restoration and regeneration, including applications in hard tissue, pulp-dentin complexes, periodontal structures, and bone. They support hydroxyapatite formation, stimulate stem cell proliferation, differentiation, and mineralization, and serve as scaffolds for cell migration and aggregation. To enhance their regenerative potential, researchers incorporate bioactive ions, polymers, and proteins—thereby improving both biological performance and physicochemical properties. Selected studies on these materials are summarized in Table 4.5.

4.4 Mechanical and Biological Compatibility of Composites

In designing composite biomaterials for biomedical use, both mechanical and biological compatibility are essential. Mechanical compatibility refers to how well a material's structural properties—such as tensile strength, elasticity, and durability—match the biomechanical demands of the target tissue. Biological compatibility involves the material's ability to function within the physiological environment without provoking adverse immune responses, while promoting cell adhesion, proliferation, and integration. Selecting a suitable biocomposite requires carefully balancing mechanical integrity with biological functionality to meet the specific clinical requirements (Table 4.6).

4.5 Innovations in Metal-Based Biomaterials

Advancements in this field focus on improving the functional performance, durability, and reliability of implants, prosthetic devices, and other biomedical systems. Current research highlights the development of next-generation metallic alloys, innovative surface modification techniques, and biofunctional coatings. These approaches aim to enhance integration between implanted materials and host tissues while minimizing postoperative complications such as microbial infections and *in vivo* degradation.

4.5.1 New Alloys and Surface Modifications

The development of novel metallic alloys has gained significant attention in the biomaterials field due to improvements in mechanical strength, flexibility (Narushima, 2019), and wear resistance (Hussein et al., 2015; Li et al., 2021). Commonly used systems include titanium-based alloys (Wang et al., 2023), cobalt–chromium formulations (Narayanan and Anand, 2022), and stainless steel—all widely applied in orthopedic (Naville et al., 2023), dental (Saha and Roy, 2023), and cardiovascular implants (Hasirci et al., 2024). Recent advances spotlight titanium–niobium alloys (Karakurt et al., 2024) for their enhanced biocompatibility, along with shape memory alloys utilized in stents (Duan et al., 2024) and bone repair devices (Meng et al., 2023). Equally important are

Table 4.5: Selected studies on ceramics and composite materials for dental regeneration.

Composition	**Role**	**Application area**	**References**
Bioglass + zinc particles	Regenerates pulp-dentin structure	Bioactive dental cement	Zhang et al., 2015
Hydroxyapatite + Strontium (Sr) particles			Jayasree et al., 2017
Zinc doped Magnesium based composites		Dental applications	Bakopoulou et al., 2016
Calcium silicate + Silver + Zinc particles			Fan et al., 2014
Boron-modified bioactive glass			Rad et al., 2019
Calcium phosphate + Gold particles			Xia et al., 2018
Magnesium-based bioceramics		Bioceramic scaffold	Theocharidou et al., 2017
Calcium strontium silicate		Supports hDPSC differentiation	Abdalla et al., 2022
Amorphous Calcium Phospate (ACP)	Promotes remineralization; facilitates enamel cell adhesion	Remineralization	Saber et al., 2011
ACP + CPP (Casein Phophopeptides) (Bioactive Protein)			Reynolds et al., 1997; Wang et al., 2024
HA + Silica filled acrylate resin		Used as a restorative resin	Chadda et al., 2016
HA + Sr particles (Bioactive Ion)		Enhances remineralization and restoration	Krishnan et al., 2016
ACP + Ag particles (Bioactive Ion)		Adds antibacterial properties	Melo et al., 2013
Ca-Si-Mg cement	Promotes PDLCs differentiation into odontogenic/ angiogenic lineages	Periodontal regeneration	Chen et al., 2016
β-TCP + Chitosan		Periodontal regeneration	Liao et al., 2010
Calsium Silicate + PCL	Bone regeneration	Bone tissue engineering, scaffolds	Lin et al., 2017
Bioglass + Li particles		Enhances osteogenesis	El Kady et al., 2016
Li-doped Calcium Phoshate Ceramics		Scaffold for tissue formation	Yuan et al., 2019
Composite Hydrogel with protein-2		Enhances osteogenesis	Xin et al., 2020

Table 4.6: Comparison of key biocomposite materials.

Composite Type	Matrix Material	Reinforcement Material	Mechanical Compatibility	Biological Compatibility	Application Area
Polymer-Matrix Composites (PMCs)	Polymers (e.g., PLA, PCL, PMMA)	Ceramic particles, fibers	Moderate strength, highly flexible	Biodegradable, bioactive with fillers	Bone scaffolds, dental implants
Ceramic-Matrix Composites (CMCs)	Ceramics (e.g., hydroxyapatite)	Glass, carbon fibers	Strong under compression, brittle	Excellent biocompatibility, non-degradable	Bone implants, dental restorations
Metal-Matrix Composites (MMCs)	Metals (e.g., titanium)	Ceramic particles, fibers	High strength, good toughness	Good biocompatibility, corrosion risk	Orthopedic implants, dental prosthetics
Natural Polymer Composites	Collagen, chitosan, alginate	Hydroxyapatite, bio-glass	Low mechanical strength, flexible	Biodegradable, highly bioactive	Tissue engineering, wound healing

surface modification strategies that improve the biological interface of metal implants. Methods such as plasma spraying (Udduttula et al., 2022), chemical etching (Aufa et al., 2022), and anodization (Shrivas et al., 2024) create micro- and nanoscale textures that support osteoblast adhesion and tissue integration (Sharma et al., 2024). Notable among these are bioactive surface coatings—such as hydroxyapatite layers (Mysore et al., 2024) and antimicrobial functionalizations (Li et al., 2023)—which reduce infection risk and improve long-term implant stability and clinical performance.

4.5.2 Biological Corrosion Resistance and Implants

Corrosion resistance is a critical requirement for metal-based implants, as corrosion can lead to the release of cytotoxic metal ions into surrounding tissues, potentially causing implant failure and adverse physiological effects (Eliaz, 2019). Biocorrosion—also known as biological corrosion—arises from electrochemical interactions between the implant surface and bodily fluids, gradually degrading the material and compromising its mechanical stability. In some cases, this may result in structural failure. Microorganisms, particularly sulfate-reducing bacteria (SRB), participate in electron transfer processes that accelerate biocorrosion by facilitating electron exchange at the implant–environment interface (Xu et al., 2023). Additionally, microbial metabolic by-products can cause local pH shifts by releasing acids or depleting alkaline compounds, which further intensify corrosion rates (Deng et al., 2022). Advances in biomaterials engineering have produced corrosion-resistant alloys such as titanium, known for its passivating oxide layer that limits ion leaching (Verma et al., 2020; Peng et al., 2022). Moreover, surface modification techniques—including diamond-like carbon (DLC) coatings (Borges et al., 2023), ceramic films (Yetim et al., 2022), and biocompatible polymer layers (Asadi et al., 2021)—have been effectively employed to enhance corrosion resistance while preserving biocompatibility. Together, these innovations help reduce degradation, improve long-term mechanical performance, and increase the clinical reliability of implants across orthopedic, dental, and cardiovascular applications.

4.6 Additive Manufacturing and Bioprinting Technologies

Advancements in additive manufacturing and bioprinting technologies have significantly transformed biomedical engineering by enabling the fabrication of patient-specific medical devices, implants, and biologically functional tissue constructs. A comprehensive understanding of the underlying technologies is essential for optimizing their clinical and research applications. Fused Deposition Modelling (FDM) offers a cost-effective and accessible platform; however, its limited resolution and material compatibility restrict its ability to produce detailed biological architectures. In contrast, Stereolithography (SLA) achieves high resolution and dimensional precision, making it suitable for complex tissue scaffolds—though potential cytotoxicity from photoinitiators in SLA resins may affect cell viability. Selective Laser Sintering (SLS) supports a wide range of biomaterials and requires no support structures, facilitating intricate geometries, yet its high-energy laser may damage heat-sensitive materials. Each bioprinting technique presents unique advantages and limitations, necessitating careful selection based on material properties and functional requirements. Collectively, these technologies have greatly expanded possibilities in regenerative medicine and tissue engineering. Most notably, they have revolutionized personalized healthcare by enabling the precise reproduction of tissue-specific microarchitectures, directly improving patient outcomes (Shanmugam et al., 2023; Kennedy et al., 2024).

4.6.1 Personalized Medicine and Implants

One of the most impactful applications of additive manufacturing in healthcare is the advancement of personalized medicine, where medical devices and implants are custom-fabricated to match an individual patient's anatomical and physiological profile . Traditional implants are typically produced in standardized sizes, often requiring intraoperative modification to accommodate patient-specific variations (Wong et al., 2018). Additive manufacturing overcomes this limitation by generating implants directly from high-resolution imaging data—such as CT and MRI scans—redefining the future trajectory of medical device fabrication. Noteworthy applications include custom-fit prosthetic limbs (Paxton et al., 2022), dental implants (Hadad et al., 2022), and orthopedic devices such as personalized hip and knee prostheses (Li et al., 2023). This patient-specific approach enhances surgical efficiency, reduces operative time, improves implant conformity, and accelerates post-operative recovery. In craniofacial reconstruction, 3D-printed implants restore bone geometry and facial symmetry with high accuracy (Slavin et al., 2023). Furthermore, porous implant architectures promote cellular adhesion and facilitate bone integration. Recent developments in bioresorbable implants—designed to degrade after tissue regeneration—eliminate the need for secondary surgeries (Su et al., 2022). The incorporation of materials like biodegradable polymers and shape-memory alloys adds functionality, enabling responsiveness to stimuli such as temperature changes and mechanical stress (Wang et al., 2021).

4.6.2 Bioinks and their Applications in Tissue Engineering

Bioprinting marks a major leap beyond traditional additive manufacturing by employing bioinks—formulations containing living cells, biomolecules, and hydrogels—to fabricate structures that closely replicate native tissue architecture (Tripathi et al., 2024). These bioinks are engineered to mimic the extracellular matrix (ECM) by integrating materials that offer both mechanical support and the essential biochemical cues for cellular viability, proliferation, and differentiation. Natural polymers such as alginate (Sangkert et al., 2021), collagen (Dzobo et al., 2019), and gelatin (Curti et al., 2022) are frequently utilized to reproduce ECM characteristics. Bioinks are typically classified according to their base materials, including natural polymers like alginate, collagen, and gelatin, and synthetic alternatives such as polyethylene glycol (PEG). Synthetic polymers such as polyvinyl alcohol (PVA), polylactic acid (PLA), and polycaprolactone (PCL) are also widely studied due to their tunable mechanics, biodegradability, and functionalizability (Cui et al., 2020; Chae and Cho, 2023). These synthetic components often form the structural backbone in hybrid bioinks, which are formulated to ensure optimal printability, mechanical fidelity, and biocompatibility (Shin et al., 2018). In regenerative medicine, such bioinks have enabled the creation of bioprinted skin grafts for burn treatment (Beheshtizadeh et al., 2020), cartilage constructs for joint repair (Sahranavard et al., 2022), and complex tissue models simulating renal and hepatic environments (Kumar and Malviya, 2024). A central research challenge remains the development of vascularized tissues, as adequate perfusion is essential for long-term cell survival in bioprinted constructs. Bioinks supplemented with stem cells and growth factors are being actively investigated for their potential to accelerate tissue repair and promote regenerative outcomes (Wang et al., 2022). Looking ahead, bioprinting holds the transformative promise of fabricating fully functional transplantable organs, helping to address the global shortage of donor tissues. With continued progress in bioink chemistry, printing precision, and post-print tissue maturation, this once-futuristic vision is rapidly becoming a clinical reality.

Table 4.7: Comparative overview of key additive manufacturing technologies with clinical applications.

Technology	Advantages	Limitations	Clinical outcomes
Fused Deposition Modelling (FDM)	Cost-effective, user-friendly, suitable for thermoplastics	Limited resolution, restricted material compatibility	Improved surgical planning and reduced operation time in craniofacial procedures (Khorsandi et al., 2021)
Stereolithography (SLA)	High resolution, smooth surface finish	Potential cytotoxicity of photoinitiators; limited bioink options	Enhanced fit and patient comfort in dental prosthetics (Danila et al., 2025)
Selective Laser Sintering (SLS)	No need for support structures; compatible with diverse biomaterials	High-energy may damage heat-sensitive components	Achieved load-bearing success in orthopedic implant applications (Fang et al., 2022)

4.7 Sustainability and Eco-Friendly Approaches in Biomaterials

With growing environmental concerns, sustainability and eco-friendliness have become key priorities in biomaterial innovation. The objective is to develop functional, biocompatible materials while minimizing environmental impact. Key areas of focus in this context include (Table 4.8):

- ***Sustainable Raw Material Sourcing*:** Biomaterials should be derived from renewable and sustainable sources to conserve natural resources.
- ***Green Manufacturing Techniques*:** Environmentally conscious production methods aim to reduce energy consumption, minimize waste, and eliminate the use of toxic chemicals.

4.8 Future Trends and Challenges

The future of biomaterials lies in developing multifunctional, biocompatible materials that enable targeted drug delivery, tissue regeneration, and personalized implants. Innovations in smart, responsive systems—triggered by pH, temperature, or magnetic fields—alongside nanotechnology and biomimetic designs, are revolutionizing regenerative medicine. Additive manufacturing and bioprinting further support custom, patient-specific solutions. However, challenges such as long-term biostability, complex regulatory pathways, high manufacturing costs, and ethical concerns must be addressed. Advancing these technologies will depend on interdisciplinary collaboration across science, engineering, medicine, and policy.

4.8.1 New Horizons in Regenerative Medicine

Regenerative medicine has advanced significantly through innovations in biomaterials, enabling effective tissue repair and organ restoration. Bioengineered scaffolds mimicking the extracellular matrix support heart valve, cartilage, and organ regeneration by promoting cell attachment and growth. Biomaterials also enhance stem cell therapies by guiding differentiation and integration, with hydrogels and nanofibers improving outcomes for conditions like diabetes and spinal injuries. Additive manufacturing has further transformed the field, enabling personalized implants and vascularized tissue constructs using bioinks that combine living cells with supportive matrices—offering new hope in addressing donor shortages and advancing precision therapies. There is increasing interest in smart biomaterials that respond to stimuli like temperature, pH, or light to release drugs, growth factors, or signals for tissue repair. Current research explores their use in wound healing, bone regeneration, and cardiovascular

Table 4.8: Sustainability and eco-friendly approaches in biomaterials.

Approach	Subtopic	Example	Short explanation & application area
Sustainable Raw Material	Plant based polymers	Cellulose	Derived from wood, cotton, and plant waste; used in bioplastics and medical dressings.
		Starch	Extracted from corn, potatoes, and cassava; used in biodegradable polymers and drug capsules.
		Alginate	Sourced from seaweed, applied in wound dressings and drug delivery systems.
		Natural rubber	Harvested from rubber trees; used in flexible biomaterials like cardiac patches.
	Marine derived materials	Chitosan	Extracted from crustacean shells; employed in wound healing and tissue engineering.
		Fish-scale collagen	Applied in skin grafts, scaffolds, and cosmetics.
		Seaweed extracts (e.g., agar)	Applied in pharmaceutical carriers and biodegradable packaging.
	Agricultural and food industry by-products	Corn and sugarcane by products	Used to produce PLA for surgical implants and biodegradable plastics.
		Fruit and vegetable waste	Converted into bioactive compounds and polymer precursors; used in antioxidant wound dressings.
		Wheat straw and rice husk	Incorporated into bio-composites and orthopedic materials.
	Protein-based material	Keratin	Derived from wool/feathers; used in wound healing and skin care.
		Silk fibroin	Extracted from silkworms; used in sutures and wound dressings, and drug delivery systems.
		Whey protein	Employed in scaffolds and bioactive packaging; used in bone engineering.
		Soy Protein	Applied in biodegradable films and wound dressings.
	Microbial biomass	Microalgae derivatives	Used in bioactive coatings and pharmaceutical applications.
		Bacterial cellulose	Fermented product, used in artificial skin and wound care.
		Fungal mycelium	Used in bio-implants and sustainable packaging.

Table 4.8 contd. ...

...Table 4.8 contd.

Approach	Subtopic	Example	Short explanation & application area
	Forestry/agroforestry resources	Cork and lignin	Used in biodegradable drug carriers and polymer systems.
		Wood and bamboo fillers	Applied in prosthetics and bio-composites.
Turing	Green synthesis techniques	Solvent-free processing	Uses melt blending or reactive extrusion; eliminates toxic solvents in polymer synthesis.
		Water-based processes	Employ water as a solvent in for hydrogels and biopolymers.
		Enzymatic polymerization	Enzyme catalyzed synthesis reduces energy and chemical usage; used in drug delivery materials.
	Bioprinting	3D-bioprinting	Minimizes material waste and enables patient-specific implant production.
		4D-bioprinting	Enables smart biomaterials to adapt to stimuli; reduces device replacement frequency.
	Biomass-derived material processing	Renewable-source extraction	Taps cellulose, chitin, starch from natural origins for dressing polymers.
		Microbial fermentation	Generates PHAs and other biodegradable materials from renewable feedstocks.
	Energy-efficient manufacturing	Low-temperature curing	Lowers thermal requirements for the polymer curing; useful in dental materials.
		Solar and bioenergy integration	Uses renewable energy in biomaterial production; supports eco-conscious medical manufacturing.

therapies, offering dynamic and tailored treatment strategies that advance regenerative medicine.

4.8.2 Ethical and Legal Dimensions of Biomaterial Innovations

As biomaterial innovations accelerate, ethical and legal challenges must be carefully addressed. Key concerns include ensuring informed consent, protecting patient rights in experimental treatments, and maintaining transparency to uphold autonomy. Intellectual property disputes—common in interdisciplinary research—can delay commercialization, highlighting the need for fair legal frameworks that protect innovation and enable public access. Cost and equitable availability remain pressing issues, as advanced therapies may be limited to wealthier populations. Ethical sourcing of human- or animal-derived materials also raises questions of consent and sustainability. Balancing innovation with ethical oversight is essential for the responsible advancement of biomaterial technologies.

4.9 Conclusion

Biomaterials have rapidly transformed contemporary healthcare, driving innovations in tissue repair, implants, and regenerative therapies. Advances in polymers, ceramics, metals, and composites have improved outcomes in diverse clinical settings—such as soft tissue repair, bone regeneration, and personalized medicine. Emerging technologies like 3D printing, bioprinting, and nanotechnology are accelerating the development of patient-specific implants and functional tissue constructs. However, challenges remain around ethical access, long-term safety, and regulatory oversight. Addressing these issues is key to translating biomaterials breakthroughs into a more effective and equitable healthcare future.

References

Abdalla, M.M., Lung, C.Y.K., Bijle, M.N. and Yiu, C.K.Y. 2022. Physicochemical properties and inductive effect of calcium strontium silicate on the differentiation of human dental pulp stem cells for vital pulp therapies: An *in vitro* study. *Materials (Basel, Switzerland)*, 15(17): 5854. doi:10.3390/ma15175854

Aghayan, N.S., Seyfi, J. and Asadollahzadeh, M.J. 2022. Preparation and characterization of antibacterial chitosan nanocomposites loaded with cellulose/chitosan nanofibers and essential oils. *Polym. Polym. Compos.*, 30: 1–10.

Altomare, L., Bonetti, L., Campiglio, E.C., Nardo, L., Draghi, L. et al. 2018. Biopolymer-based strategies in the design of smart medical devices and artificial organs. *Int. J. Artif. Organs.*, 41(6): 337–359. doi: 10.1177/0391398818765323.

Araújo, M., Viveiros, R. and Correia, T.R. 2014. Natural melanin: A potential pH-responsive drug release device. *Int. J. Pharm.*, 469(1): 140–145.

Asadi, H., Suganthan, B., Ghalei, S., Handa, H., Ramasamy, R.P. et al. 2021. A multifunctional polymeric coating incorporating laws one with corrosion resistance and antibacterial activity for biomedical Mg alloys. *Prog. Org. Coat.*, 153–106157. https://doi.org/10.1016/j.porgcoat.2021.106157.

Bakopoulou, A., Papachristou, E. and Bousnaki, M. 2016. Human treated dentin matrices combined with Zn-doped, Mg-based bioceramic scaffolds and human dental pulp stem cells toward targeted dentin regeneration. *Dent. Mat.*, 32(8): e159–e175. doi:10.1016/j.dental.2016.05.013.

Ballios, B.G., Cooke, M.J., van der Kooy, D. et al. 2010. A hydrogel-based stem cell delivery system to treat retinal degenerative diseases. *Biomaterials*, 31(9): 2555–2564.

Bavya Devi, K., Nandi, S.K. and Roy, M. 2019. Magnesium silicate bioceramics for bone regeneration: A review. *J. Indian Inst. Sci.*, 99: 261–288.

Beheshtizadeh, N., Lotfibakhshaiesh, N. and Pazhouhnia, Z. 2020. A review of 3D bioprinting for bone and skin tissue engineering: A commercial approach. *J. Mater. Sci.*, 55: 3729–3749. https://doi.org/10.1007/s10853-019-04259-0.

Borges, G.A., Costa, R.C. and Nagay, B.E. 2023. Targeting biomechanical endurance of dental-implant abutments using a diamond-like carbon coating. *ACS Appl. Bio. Mater.*, 6(12): 5630–5643. doi: 10.1021/acsabm.3c00802.

Cao, H. and Kuboyama, N. 2010. A biodegradable porous composite scaffold of PGA/beta-TCP for bone tissue engineering. *Bone*, 46: 386–395.

Cao, W. and Hench, L.L. 1996. Bioactive materials. *Ceram. Int.*, 22: 493–507.

Chadda, H., Naveen, S.V., Mohan, S., Satapathy, B.K., Ray, A.R. and Kamarul, T. 2016. Cytotoxic evaluation of hydroxy apatite-filled and silica/hydroxy apatite-filled acrylate-based rest orative composite resins: An *in vitro* study. *J. Prosth. Dent.*, 116(1): 129–35.

Chae, S. and Cho, D.W. 2023. Biomaterial-based 3D bioprinting strategy for orthopedic tissue engineering. *Acta Biomaterialia*, 156: 4–20. doi:10.1016/j.actbio.2022.08.004.

Chen, C.H., Tsai, C.C. and Chen, W. 2006. Novel living cell sheet harvest system composed of thermoreversible methylcellulose hydrogels. *Biomacromolecules*, 7(3): 736–743.

Chen, Y.W., Hsu, T.T., Wang, K. and Shie, M.Y. 2016. Preparation of fast-setting and degradable Ca-Si-Mg cement with both odontogenesis and angiogenesis differentiation of human periodontal ligament cells. *Mater. Sci. Eng. C Mater. Biol. Appl.*, 60: 374–383. doi:10.1016/j.msec.2015.11.064.

Chiu, Y.-L., Chen, M.-C. and Chen, C.-Y. 2009. Rapidly *in situ* forming hydrophobically-modified chitosan hydrogels via pH-responsive nanostructure transformation. *Soft Matter*, 5(5): 962–965.

Contessi, N., Altomare, L. and Filipponi, A. 2017. Thermo-responsive properties of methylcellulose hydrogels for cell sheet engineering. *Mater. Lett.*, 207: 157–160.

Cui, X., Li, J. and Hartanto, Y. 2020. Advances in extrusion 3D bioprinting: A focus on multicomponent hydrogel-based bioinks. *Adv. Healthcare Mater.*, 15(9): 1901648. https://doi.org/10.1002/adhm.201901648.

Curti, F. and Serafim, A. 2022. Development of biocomposite alginate-cuttlebone-gelatin 3D printing inks designed for scaffolds with bone regeneration potential. *J. Biomater. Appl. Mar. Drugs*, 20(11): 670. https://doi.org/10.3390/md20110670.

Danila, A.I., Breban-Schwarzkopf, D., Daescu, E., Olariu, I., Dinu, S. et al. 2025. Systematic review of the quality of stereolithographic three-dimensionally printed materials for provisional dental restorations. *Materials (Basel)*, 18(3): 721. doi: 10.3390/ma18030721.

De Aza, A., Chevalier, J., Fantozzi, G., Schehl, M., Torrecillas, R. et al. 2002. Crack growth resistance of alumina, zirconia and zirconia toughened alumina ceramics for joint prostheses. *Biomaterials*, 23: 937–945.

De Nardo, L., Bertoldi, S. and Tanzi, M.C. 2011. Shape memory polymer cellular solid design for medical applications. *Smart Mater. Struct.*, 20(3): 035004.

De Nardo, L., Moscatelli, M. and Silvi, F. 2010. Chemico-physical modifications induced by plasma and ozone sterilizations on shape memory polyurethane foams. *J. Mater. Sci. Mater. Med.*, 21(7): 2067–2078.

Deng, X., Gould, M. and Ali, M.A. 2022. A review of current advancements for wound healing: Biomaterial applications and medical devices. *J. Biomed. Mater. Res.*, 110(11): 2542–2573. doi:10.1002/jbm.b.35086.

Dong, C. and Lv, Y. 2016. Application of collagen scaffold in tissue engineering: Recent advances and new perspectives. *Polymers*, 8: 42. https://doi.org/10.3390/polym8020042.

Dong, Z., Wang, Q. and Du, Y. 2006. Alginate/gelatin blend films and their properties for drug controlled release. *J. Memb. Sci.*, 280(1-2): 37–44.

Duan, X. 2024. Research progress of metal biomaterials with potential applications as cardiovascular stents and their surface treatment methods to improve biocompatibility. *Heliyon*, 10(4): e25515.

Dzobo, K., Motaung, K.S.C.M. and Adeside, A. 2019. Recent trends in decellularized extracellular matrix bioinks for 3D printing: An updated review. *Mater. Adv. Int. J. Mol. Sci.*, 18: 20(18)–4628. doi: 10.3390/ijms20184628.

El-Kady, A.M., Farag, M.M. and El-Rashedi, A.M. 2016. Bioactive glass nanoparticles designed for multiple deliveries of lithium ions and drugs: Curative and restorative bone treatment. *Eur. J. Pharm. Sci.*, 91: 243–250. doi:10.1016/j .ejps .2016.05.004.

Eliaz, N. 2019. Corrosion of metallic biomaterials: A review. *Mater.*, 12(3): 407. https://doi.org/10.3390/ma12030407.

Erten, E. and Arslan, Y.E. 2018. The great harmony in translational medicine: Biomaterials and stem cells. *Adv. Exp. Med. Biol.*, 1119: 21–39.

Farag, M.M. 2023. Recent trends on biomaterials for tissue regeneration applications: Review. *J. Mater. Sci.*, 58: 527–558. doi: 10.1007/s10853-022-08102-x.

Fan, W., Wu, D., Tay, F.R., Ma, T., Wu, Y. et al. 2014. Effects of adsorbed and templated nanosilver in mesoporous calcium-silicate nanoparticles on inhibition of bacteria colonization of dentin. *Inter. J. Nanomed.*, 9: 5217–5230. doi:10.2147/ijn.S73144.

Fang, S., Wang, Y., Xu, P., Zhu, J., Liu, J. et al. 2022. Three-dimensional-printed titanium implants for severe acetabular bone defects in revision hip arthroplasty: Short- and mid-term results. *Int. Orthop.*, 46(6): 1289–1297. doi: 10.1007/s00264-022-05390-5.

Fattahi, F.S. 2020. Poly (lactic acid) nano-structures for cartilage regeneration and joint repair: Strategies and ideas. *Recent Trends, Nanosci. Technol.*, 1: 1–19.

Fattahi, F.S., Khoddami, A. and Avinc, O. 2019. Poly (lactic acid) (PLA) nanofibers for bone tissue engineering. *J. Text. Polym.*, 7: 47–64.

Gialouri, A., Saravanou, S.F., Loukelis, K., Chatzinikolaidou, M., Pasparakis, G. et al. 2023. Thermoresponsive alginate-Graft-pNIPAM/Methyl cellulose 3D-printed scaffolds promote osteogenesis *in vitro*. *Gels*, 9: 984. https://doi.org/10.3390/gels9120984.

Guillén-Carvajal, K., Valdez-Salas, B. and Beltrán-Partida, E. 2023. Chitosan, gelatin, and collagen hydrogels for bone regeneration. *Polymers*, 15(13): 2762. doi:10.3390/polym15132762.

Guo, J. and Kaletunç, G. 2016. Dissolution kinetics of pH responsive alginate-pectin hydrogel particles. *Food Res. Int.*, 88: 129–139.

Guo, D., Mao, M., Qi, W., Li, H., Ni, P. et al. 2011. The influence of Sr and H_3PO_4 concentration on the hydration of SrCaHA bone cement. *J. Mater. Sci. Mater. Med.*, 22: 2631–2640.

Guo, D., Xu, K., Zhao, X. and Han, Y. 2005. Development of a strontium–containing hydroxyapatite bone cement. *Biomaterials*, 26: 4073–4083.

Gupta, D., Tator, C.H. and Shoichet, M.S. 2006. Fast-gelling injectable blend of hyaluronan and methylcellulose for intrathecal, localized delivery to the injured spinal cord. *Biomaterials*, 27(11): 2370–2379.

Hadad, H., Lima, F.B.B. and Shirinbak, I. 2023. The impact of 3D printing on oral and maxillofacial surgery. *J. 3D Print. Med.*, 7(2). https://doi.org/10.2217/3dp-2022-0025.

Hasirci, V. and Hasirci, N. 2024. Metal implants. In: *Fundamentals of Biomaterials*. Springer, Cham. https://doi.org/10.1007/978-3-031-54046-2_3.

Heimann, R.B. 2013. Structure, properties, and biomedical performance of osteoconductive bioceramic coatings. *Surf. Coat. Technol.*, 233: 27–38.

Heness, G. and Ben-Nissan, B. 2004. Innovative bioceramics. *Mater. Forum*, 27: 104–114.

Huang, Y.X., Ren, J., Chen, C., Ren, T.B., Zhou, X.Y. et al. 2008. Preparation and properties of poly(lactide-co-glycolide) (PLGA)/nanohydroxyapatite (NHA) scaffolds by thermally induced phase separation and rabbit MSCs culture on scaffolds. *J. Biomater. Appl.*, 22: 409–432.

Hussain, T.M.M., Patil, A.Y., Hegde, C., Sudeept, M.A., Kumar, R. et al. 2024. Apatite insights: From synthesis to biomedical applications. *Eur. Polym. J.*, 209: 112842. ISSN: 0014-3057. https://doi.org/10.1016/j.eurpolymj.2024.112842.

Hussein, M.A. and Mohammed, A.S. 2015. Wear characteristics of metallic biomaterials: A review. *Mater. (Basel).*, 5: 2749–2768. doi: 10.3390/ma8052749.

Jaidev, L.R. and Chatterjee, K. 2019. Surface functionalization of 3D printed polymer scaffolds to augment stem cell response. *Mater. Des.*, 161: 44–54.

Jayasree, R., Kumar, T.S.S., Mahalaxmi, S., Abburi, S., Rubaiya, Y. and Doble, M. 2017. Dentin remineralizing ability and enhanced antibacterial activity of strontium and hydroxylion co-releasing radiopaque hydroxy apatite cement. *J. Mater. Sci. Mater. Med.*, 28(6): 95. doi:10.1007/s10856-017-5903- x.

Jiale Wang, Linglin Zhang and Kun Wang. 2024. Bioactive ceramic-based materials: Beneficial properties and potential applications in dental repair and regeneration. *Regenerative Medicine*, 19: 5, 257–278. doi: 10.1080/17460751.2024.2343555.

Jindal, A., Mondal, T. and Bhattacharya, J. 2020. An *in vitro* evaluation of zinc silicate fortified chitosan scaffolds for bone tissue engineering. *Int. J. Biol. Macromol.*, 164: 4254–4262. doi: 10.1016/j.ijbiomac.2020.09.018.

Kang, C., Poon, P. and Tator, C. 2009. A new paradigm for local and sustained release of therapeutic molecules to the injured spinal cord for neuroprotection and tissue repair. *Tissue Eng. Part A*, 15(3): 595–604.

Karakurt, E.M., Huang, Y. and Cetin, Y. 2024. Assessing microstructural, biomechanical, and biocompatible properties of TiNb alloys for potential use as load-bearing implants. *J. Funct. Biomater.*, 15(9): 253. https://doi.org/10.3390/jfb15090253.

Kennedy, S.M., K A, J.J.J., et al. 2024. Transformative applications of additive manufacturing in biomedical engineering: Bioprinting to surgical innovations. *J. Med. Eng. Technol.*, 48(4): 151–168. https://doi.org/10.1080/03091902.2024.2399017.

Khorsandi, D., Fahimipour, A., Abasian, P., Saber, S.S., Seyedi, M. et al. 2021. 3D and 4D printing in dentistry and maxillofacial surgery: Printing techniques, materials, and applications. *Acta Biomater.*, 122: 26–49. doi: 10.1016/j.actbio.2020.12.044.

Koumentakou, I., Michopoulou, A. and Noordam, M.J. 2024. 3D-printed scaffolds for tissue engineering applications using thermosensitive hydrogels based on biopolymer blends. *J. Mater. Sci.*, 59; 9021–9041. https://doi.org/10.1007/s10853-024-09707-0.

Kumar, S. and Malviya, R. 2024. Bioprinting of hepatic tissue using 3D technology: Transitioning beyond laboratory models to real-world applications in medical treatments. *Appl. Mater.Today*, 39: 102307. https://doi.org/10.1016/j.apmt.2024.102307.

Leng, D., Li, Y. and Zhu, J. 2020. The antibiofilm activity and mechanism of nanosilver- and nano zinc-incorporated mesoporous calcium-silicate nanoparticles. *Int. J. Nanomed.*, 15: 3921–3936. doi:10.2147/ijn.S244686.

Li, C., Guo, C. and Fitzpatrick, V. 2020. Design of biodegradable, implantable devices towards clinical translation. *Nat. Rev. Mater.*, 5: 61–81. https://doi.org/10.1038/s41578-019-0150-z.

Li, H.F., Huang, J.Y. and Lin, G.C. 2021. Recent advances in tribological and wear properties of biomedical metallic materials. *Rare Met.*, 40: 3091–3106. https://doi.org/10.1007/s12598-021-01796-z.

Li, X. and Coates, D.E. (ed.). 2023. Hollow channels scaffold in bone regenerative: A review. *J. Biomater. Sci. Polym.*, 34(12): 1702–1715. https://doi.org/10.1080/09205063.2023.2181066.

Liao, F., Chen, Y. and Li, Z. 2010. A novel bioactive three-dimensional beta-tricalcium phosphate/chitosan scaffold for periodontal tissue engineering. *J. Mater. Sci. Mater. Med.*, 21(2): 489–496.

Lin, Y.H., Chiu, Y.C., Shen, Y.F., Wu, Y.A., Shie, M.Y. et al. 2017. Bioactive calcium silicate/poly-caprolactone composite scaffolds 3D printed under mild conditions for bone tissue engineering. *J. Mater. Sci. Mater. Med.*, 29(1): 11. doi:10.1007/s10856-017-6020-6.

Maity, S., Mukhopadhyay, P. and Kundu, P.P. 2017. Alginate coated chitosan core-shell nanoparticles for efficient oral delivery of naringenin in diabetic animals in an *in vitro* and *in vivo* approach. *Carbohydr. Polym.*, 170: 124–132.

Marin, E. 2023. History of dental biomaterials: Biocompatibility, durability and still open challenges. *Herit. Sci.*, 11: 207. https://doi.org/10.1186/s40494-023-01046-8.

Mazzocchi, M. and Bellosi, A. 2008. On the possibility of silicon nitride as a ceramic for structural orthopaedic implants. Part I: Processing, microstructure, mechanical properties, cytotoxicity. *J. Mater. Sci. Mater. Med.*, 19: 2881–2887.

Melo, M.A., Cheng, L., Zhang, K., Weir, M.D., Rodrigues, L.K. et al. 2013. Novel dental adhesives containing nanoparticles of silver and amorphous calcium phosphate. *Dent. Mater.*, 29(2): 199–210.

Melone, L., Altomare, L. and Cigada, A. 2012. Phase change material cellulosic composites for the cold storage of perishable products: from material preparation to computational evaluation. *Appl. Energ.*, 89(1): 339–346.

Meng, M., Wang, J. and Huang, H. 2023. 3D printing metal implants in orthopedic surgery: Methods, applications and future prospects. *J. Orthop. Transl.*, 42: 94–112. https://doi.org/10.1016/j.jot.2023.08.004.

Moonesi Rad, R., Pazarçeviren, E. and Ece Akgün, E. 2019. *In vitro* performance of a nanobiocomposite scaffold containing boron-modified bioactive glass nanoparticles for dentin regeneration. *J. Biomater. Appl.*, 33(6): 834–853. doi:10.1177/0885328218812487.

Narushima, T. 2019. New-generation metallic biomaterials. pp. 495–521. *In*: Niinomi, M. (ed.). *Metals for Biomedical Devices* (Second Edition), Woodhead Publishing Series in Biomaterials, Woodhead Publishing. ISBN: 9780081026663. https://doi.org/10.1016/B978-0-08-102666-3.00019-5.

Naville, W., Magnabosco, R. and Costa, I. 2023. Uniaxial plastic strain effect on the corrosion-fatigue resistance of ISO 5832-1 stainless steel biomaterial. *Int. J. Fatigue*, 173: 107701. https://doi.org/10.1016/j.ijfatigue.2023.107701.

Olad, A. and Hagh, H.B.K. 2019. Graphene oxide and amin-modified graphene oxide incorporated chitosan-gelatin scaffolds as promising materials for tissue engineering. *Compos. Part. B Eng.*, 162: 692–702. doi: 10.1016/j.compositesb.2019.01.040.

Olofsson, J., Pettersson, M., Teuscher, N., Heilmann, A., Larsson, K. et al. 2012. Fabrication and evaluation of SixNy coatings for total joint replacements. *J. Mater. Sci. Mater. Med.*, 23: 1879–1889.

Paxton, N.C., Nightingale, R.C. and Woodruff, M.A. 2022. Capturing patient anatomy for designing and manufacturing personalized prostheses. *Curr. Opin. Biotechnol.*, 73: 282–289. https://doi.org/10.1016/j.copbio.2021.09.004.

Peng, Y., Peng, J., Wang, Z., Xiao, Y. and Qiu, X. 2022. Diamond-like carbon coatings in the biomedical field: Properties, applications and future development. *Coatings*, 12: 1088. https://doi.org/10.3390/coatings12081088.

Qian, J., Xu, W., Yong, X. and Zhang, W. 2014. Fabrication and *in vitro* biocompatibility of biomorphic PLGA/nHA composite scaffolds for bone tissue engineering. *Mater. Sci. Eng. C. Mater. Biol. Appl.*, 36: 95–101.

Saber-Samandari, S. and Gross, K.A. 2011. Amorphous calcium phosphate offers improved crack resistance: A design feature from nature? *Acta Biomater.*, 7(12): 4235–4241. doi:10.1016/j.actbio.2011.06.048.

Saha, S. and Roy, S. 2022. Metallic dental implants wear mechanisms, materials, and manufacturing processes: A literature review. *Mater. (Basel).*, 16(1): 161. doi: 10.3390/ma16010161. PMID: 36614500; PMCID: PMC9821388.

Sahranavard, M., Sarkari, S. and Safavi, S. 2022. Three-dimensional bioprinting of decellularized extracellular matrix-based bioinks for cartilage regeneration: A systematic review. *Biomater. Transl.*, 3(2): 105–115. https://doi.org/10.12336/biomatertransl.2022.02.004.

Saito, H., Taguchi, T. and Aoki, H. 2007. pH-responsive swelling behavior of collagen gels prepared by novel crosslinkers based on naturally derived di- or tricarboxylic acids. *Acta Biomater.*, 3(1): 89–94.

Sangkert, S., Kamolmatyakul, S. and Gelinsky, M. 2021. 3D printed scaffolds of alginate/polyvinyl alcohol with silk fibroin based on mimicked extracellular matrix for bone tissue engineering in maxillofacial surgery. *Mater. Today Commun.*, 26: 102140. https://doi.org/10.1016/j.mtcomm.2021.102140.

Satchanska, G., Davidova, S. and Petrov, P.D. 2024. Natural and synthetic polymers for biomedical and environmental applications. *Polymers*, 16(8): 1159. doi: 10.3390/polym16081159.

Schneider, K.H., Goldberg, B.J., Hasturk, O., Mu, X., Dötzlhofer, M. et al. 2023. Silk fibroin, gelatin, and human placenta extracellular matrix-based composite hydrogels for 3D bioprinting and soft tissue engineering. *Biomater. Res.*, 27: 117. doi: 10.1186/s40824-023-00431-5.

Sha, L., Chen, Z., Chen, Z., Zhang, A., Yang, Z. et al. 2016. Polylactic acid based nanocomposites: Promising safe and biodegradable materials in biomedical field. *Int. J. Polym. Sci.*, 2016: 6869154.

Shanmugam, D.K., Anitha, S.C. and Souresh, V. 2023. Current advancements in the development of bionic organs using regenerative medicine and 3D tissue engineering. *Mater. Technol.*, 38(1). https://doi.org/10.1080/10667857.2023.2242732.

Sharma, P., Mishra, V. and Murab, S. Unlocking osseointegration: Surface engineering strategies for enhanced dental implant integration. *ACS Biomater. Sci. Eng.* Article ASAP. Doi: 10.1021/acsbiomaterials.4c01178.

Shin, J.H. and Kang, H.W. 2018. The development of gelatin-based bio-ink for use in 3D hybrid bioprinting. *Int. J. Precis. Eng. Manuf.*, 19: 767–771. https://doi.org/10.1007/s12541-018-0092-1.

Shrestha, S., Yeon Lee, S., Shrestha, D., Kandel, R., Yoo, Y. et al. 2022. Micro/nanometer-sized porous structure of zinc phosphate incorporated $Ti(HPO_4)_2$ hydrate bioceramic induces osteogenic gene expression and enhances osteoporotic bone regeneration. *Chem. Eng. J.*, 450: 138360.

Shrivas, S., Samaur, H. and Yadav, V. 2024. Soft and hard tissue integration around percutaneous bone-anchored titanium prostheses: Toward achieving holistic biointegration. *ACS Biomater. Sci. Eng.*, 10(4): 1966–1987. doi10.1021/acsbiomaterials.3c01555.

Silva, A.K.A., Richard, C. and Ducouret, G. 2013. Xyloglucan-derivatized films for the culture of adherent cells and their thermocontrolled detachment: A promising alternative to cells sensitive to protease treatment. *Biomacromolecules*, 14(2): 512–519.

Slavin, B.V., Ehlen, Q.T. and Costello, J.P. 2023. 3D printing applications for craniomaxillofacial reconstruction: A sweeping review. *ACS Biomater. Sci. Eng.*, 9(12): 6586–6609. https://doi.org/10.1021/acsbiomaterials.3c01171.

Su, Y., Fu, J. and Lee, W. 2022. Improved mechanical, degradation, and biological performances of Zn–Fe alloys as bioresorbable implants. *Bioact. Mater.*, 17: 334–343. https://doi.org/10.1016/j.bioactmat.2021.12.030.

Sun, Z., Ma, L. and Sun, X. 2023. The overview of antimicrobial peptide-coated implants against oral bacterial infections. *Aggregate*, 4–e09. https://doi.org/10.1002/agt2.309.

Tanvir, M.A.H., Khaleque, M.A., Kim, G.-H., Yoo, W.-Y. and Kim, Y.-Y. 2024. The role of bioceramics for bone regeneration: History, mechanisms, and future perspectives. *Biomimetics*, 9: 230. https://doi.org/ 10.3390/biomimetics9040230.

Theocharidou, A., Bakopoulou, A. and Kontonasaki, E. 2017. Odontogenic differentiation and biomineralization potential of dental pulp stem cells inside Mg-based bioceramic scaffolds under low-level laser treatment. *Las. Med. Sci.*, 32(1): 201–210. doi:10.1007/s10103-016- 2102-9.

Tomić, S.L., Babić Radić, M.M., Vuković, J.S., Filipović, V.V., Nikodinovic-Runic, J. et al. 2023. Alginate-based hydrogels and scaffolds for biomedical applications. *Mar. Drugs*, 21: 177. https://doi.org/10.3390/md21030177.

Tripathi, S., Dash, M. and Chakraborty, R. 2020. Engineering considerations in the design of tissue-specific bioink for 3D bioprinting applications. *Biomater. Res.*, 24(1): 57.

Udduttula, A., Li, J. and Ma, Z. 2022. A novel apatite-inspired Sr5(PO4)2SiO4 plasma-sprayed coating on Ti alloy promoting biomineralization, osteogenesis, and angiogenesis. *Ceram. Int.*, 48(8): 10979–10989. ISSN: 0272-8842. https://doi.org/10.1016/j.ceramint.2021.12.317.

Verma, R.P. 2020. Titanium based biomaterial for bone implants: A mini review. *Mater. Today Proc.*, 26(2): 3148–3151. ISSN: 2214-7853. https://doi.org/10.1016/j.matpr.2020.02.649.

Wang, L., Li, Y. and Zhou, L. 2023. Progress in additive manufacturing, additive repair and fatigue evaluation of aviation titanium alloy blades. *Mater. Res. Lett.*, 11(12): 973–1012. https://doi.org /10.1080/21663831.2023.2275599.

Wang, Y., Venezuela, J. and Dargusch, M. 2021. Biodegradable shape memory alloys: Progress and prospects. *Biomaterials*, 279: 121215. https://doi.org/10.1016/j.biomaterials.2021.121215.

Wang, Y., Yuan, X. and Yao, B. 2022. Tailoring bioinks of extrusion-based bioprinting for cutaneous wound healing. *Bioact. Mater.*, 17: 178–194. https://doi.org/10.1016/j.bioactmat.2021.12.030.

Wells, M.R., Kraus, K. and Batter, D.K. 1997. Gel matrix vehicles for growth factor application in nerve gap injuries repaired with tubes: A comparison of biomatrix, collagen, and methylcellulose. *Exp. Neurol.*, 146(2): 395–402.

Wong, K.C. and Scheinemann, P. Additive manufactured metallic implants for orthopaedic applications. *Sci. China Mater*. https://doi.org/10.1007/s40843-017-9243-9.

Xia, Y., Chen, H., Zhang, F. et al. 2018. Gold nanoparticles in injectable calcium phosphate cement enhance osteogenic differentiation of human dental pulp stem cells. *Nanomed. Nanotechnol. Biol. Med.*, 14(1): 35–45. doi:10.1016/j.nano.2017.08.014.

Xin, T., Mao, J. and Liu, L. 2020. Programmed sustained release of recombinant human bone morphogenetic protein-2 and inorganic ion composite hydrogel as artificial periosteum. *ACS Appl. Mater. Interf.*, 12(6): 6840–6851. doi:10.1021/acsami.9b18496.

Xu, D., Gu, T. and Lovley, D.R. 2023. Microbially mediated metal corrosion. *Nat. Rev. Microbiol.*, 21: 705–718. doi: 10.1038/s41579-023-00920-3.

Yetim, T., Turalioğlu, K. and Taftali, M.A.F. 2022. Synthesis and characterization of wear and corrosion resistant Ni-doped Al_2O_3 nanocomposite ceramic coatings by sol-gel method. *Surf. Coat. Technol.*, 444–128659. ISSN: 0257-8972. https://doi.org/10.1016/j.surfcoat.2022.128659.

Yuan, Y., Yuan, Q. and Wu, C. 2019. Enhanced osteoconductivity and osseointegration in calcium polyphosphate bioceramic scaffold via lithium doping for bone regeneration. *ACS Biomater. Sci. Eng.*, 5(11): 5872–5880. doi:10.1021/acsbiomaterials.9b00950.

Zhang, J., Park, Y.D. and Bae, W.J. 2015. Effects of bioactive cements incorporating zinc-bioglass nanoparticles on odontogenic and angiogenic potential of human dental pulp cells. *J. Biomater. Appl.*, 29(7): 954–964. doi:10.1177/0885328214550896.

Zhang, J., Iwasa, M., Kotobuki, N., Tanaka, T., Hirose, M., Ohgushi, H. and Jiang, D. 2006. Fabrication of hydroxyapatite–zirconia composites for orthopedic applications. *J. Am. Ceram. Soc.*, 89: 3348–3355.

Zhang, X., Gubbels, G.M., Terpstra, R. and Metselaar, R. 1997. Toughening of calcium hydroxyapatite with silver particles. *J. Mater. Sci.*, 32: 235–243.

Zhang, Y., Tan, S. and Yin, Y.C. 2003. Fibre reinforced hydroxyapatite bioceramics. *Ceram. Int.*, 29: 113–116.

Zhao, X., Yang, J., Xin, H., Wang, X., Zhang, L. et al. 2018. Improved dispersion of SiC whisker in nano hydroxyapatite and effect of atmospheres on sintering of the SiC whisker reinforced nano hydroxyapatite composites. *Mater. Sci. Eng. C*, 91: 135–145.

Zheng, R., Duan, H., Xue, J., Liu, Y., Feng, B. et al. 2014. The influence of gelatin/PCL ratio and 3-D construct shape of electrospun membranes on cartilage regeneration. *Biomaterials*, 35: 152–164.

CHAPTER

5 Alginate-based Biocomposites for Environmental Remediation

Ashish Kapoor, Anjali Awasthi* and *Pranava Chaudhari*

The widespread issue of environmental pollution has driven the advancement of remediation technologies. A variety of natural and synthetic materials have been explored for developing effective techniques to remove contaminants from water. Among these, alginate-based materials stand out due to their natural origin, ease of functional integration, and biocompatibility. This chapter highlights recent progress in the synthesis and application of alginate-based biocomposites for water contaminant removal. Alginate can be combined with materials such as activated carbon, biocarbon, metallic oxides, and nanomaterials. Typically synthesized via ionic gelation, these composites form porous structures with high surface area and multiple active sites that facilitate pollutant binding through electrostatic interactions, hydrogen bonding, and ion exchange. Exploring their structure-property relationships reveals improvements in sorption capacity, selectivity and structural integrity. Their performance is critically assessed for removing diverse pollutants including heavy metals, dyes, pharmaceuticals, and biological contaminants.

Department of Chemical Engineering, Harcourt Butler Technical University, Kanpur, India.
ORCID IDs of authors: https://orcid.org/0000-0002-4686-394X; https://orcid.org/0009-0005-4813-431X; https://orcid.org/0000-0002-5701-5858
* Corresponding author: ashishk@hbtu.ac.in

DOI: 10.1201/9781003607359-5

5.1 Introduction

Environmental contamination poses a serious threat to all life forms, with water pollution being especially concerning due to rising industrial activity and inadequate pollution control. Industrial effluents and agricultural runoff—rich in chemicals, pesticides, and fertilizers—often infiltrate water and soil, eventually entering the food chain. These pollutants contribute to bioaccumulation and can cause significant ecological and health hazards. To address growing environmental challenges, several wastewater treatment technologies have been developed, including coagulation, membrane filtration, advanced oxidation, and microbial methods (Mojiri and Bashir, 2022). While effective under lab-scale conditions, these methods often face barriers such as high costs, complex implementation, and intensive energy demands, limiting large-scale application (Rashid et al., 2021). Adsorption has emerged as a promising alternative due to its simplicity, adaptability, and strong pollutant removal performance (Awasthi et al., 2023). However, its success depends on the properties of the adsorbent and operational conditions—prompting increased research into sustainable material options.

Natural biopolymers like alginate are emerging as eco-friendly solutions for water treatment (Muthukumaran et al., 2021). Extracted from brown seaweed, alginate is biodegradable, non-toxic, and abundantly available (Anjana and Arunkumar, 2024). Its ability to form hydrogels with divalent cations enables fabrication into structures such as beads and films for pollutant adsorption (Pinheiro et al., 2021; Dechojarassri et al., 2018; Mohammed et al., 2019). Functional groups like carboxyl and hydroxyl enhance contaminant binding, and hybridization with materials such as biocarbon, clays, nanomaterials, or microbes further improves performance (Zdiri et al., 2022; Doyo et al., 2023; Fernando et al., 2020). Recent advancements highlight novel synthesis techniques and applications of alginate-based biocomposites for water remediation (Khoj et al., 2024; Faouzia et al., 2025). An overview is presented in Fig. 5.1.

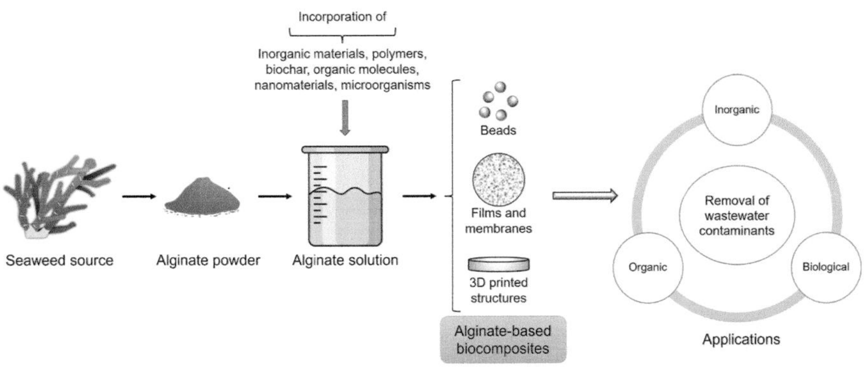

Figure 5.1: Illustrative overview of synthesis and water-treatment applications of alginate-based composites.

Recent advancements in additive manufacturing and material functionalization have led to innovative solutions for environmental remediation (Luo et al., 2024; El-Aassar et al., 2024). However, challenges remain, including limited column studies, insufficient flow stability, real-world deployment barriers, and the removal of emerging pollutants (Hastuti et al., 2025; Xin et al., 2025). Alginate-based hybrid materials offer a sustainable, green alternative to conventional adsorbents. This chapter reviews their synthesis, functionality, and environmental applications while also addressing practical challenges and future directions for field implementation.

5.2 Preparation of Alginate-based Biocomposites

Alginate is a naturally derived anionic polysaccharide found in brown algae, primarily used as sodium alginate. It consists of a linear chain of β-D-mannuronic acid (M) and α-L-guluronic acid (G) units, arranged as homopolymeric or alternating blocks. Its abundant functional groups contribute to its strong reactivity and binding capacity, making it useful in various applications. A unique property of alginate is its ability to form hydrogels upon interacting with divalent cations like calcium (Ca^{2+}). In this ion-exchange process, sodium ions (Na^{+}) are replaced, resulting in the formation of a calcium alginate gel with a thermally irreversible crosslinked structure. Sodium alginate is non-toxic, stable, and valued for its excellent gelation, film-forming, and complexation capabilities. Alginate finds broad use across sectors like agriculture, food, pharmaceuticals, cosmetics, textiles, and environmental engineering. In water treatment, its functional groups enable ion exchange with cationic and metal pollutants, making it effective for contaminant removal. Although alginate offers key benefits, its use in complex water systems is limited by low mechanical strength, structural instability, and low surface area—mainly due to its linear molecular structure. These shortcomings reduce its standalone performance in pollutant removal. However, forming composites with polymers, nanoparticles, or microbes has shown significant promise. Such modifications enhance surface area, stability, and functionality, making alginate-based sorbents more effective and adaptable for targeted environmental applications.

5.2.1 Synthesis Approaches

The synthesis of alginate-based biocomposites primarily involves two approaches: physical modifications, which incorporate nanomaterials, polymers, or macromolecules into the alginate matrix, and chemical modifications, which alter alginate's molecular structure through techniques like graft copolymerization and crosslinking to enhance strength, stability, and thermal resistance (Guo et al., 2023). These strategies help address the inherent limitations of pure alginate.

5.2.1.1 Synthesis via Physical Modifications

This approach involves blending alginate with nanomaterials, clays, polymers, or macromolecules to improve mechanical strength and surface area. The resulting biocomposites interact with contaminants through π–π stacking, hydrogen bonding, and electrostatic forces. Sorption site characteristics vary by additive; composites with natural macromolecules often show enhanced stability and sorption efficiency due to functional group interactions and alginate chain binding.

Clay materials are widely explored for their high surface area, ion-exchange capacity, low cost, and environmental friendliness—making them ideal for adsorbent development. Their incorporation into alginate has led to biocomposites with improved mechanical and sorption properties. One study reported acid-treated clay-alginate beads effectively removed hazardous dyes (Kausar et al., 2020). Another study developed alginate beads encapsulating zero-valent iron (Fe^0) and nanomaterials for nitrate removal (Cho et al., 2015). Despite slightly slower kinetics than bare Fe^0/NMT, the composite beads showed enhanced stability, prolonged activity, and reduced nitrate levels below permissible limits across multiple cycles—highlighting their suitability for groundwater remediation.

5.2.1.2 Synthesis via Chemical Modifications

Chemical modification of sodium alginate—via esterification, oxidation, crosslinking, graft copolymerization, or click reactions—enhances its hydrophilicity, binding capacity, and overall functionality. These changes improve strength, stability, and surface area while increasing sorption site density, making modified alginates more effective for wastewater treatment. A recent study fabricated sodium alginate-based graft polymer composite beads using a free-radical graft polymerization technique in a surface-crosslinked bead reactor (Wang et al., 2022). A homogeneous sodium alginate solution was prepared with ammonium persulfate (initiator), N,N′-methylene-bis-acrylamide (crosslinker), acrylamide, sepiolite, and gelatin. The mixture was dropped into a 5% $CaCl_2$ solution at 70°C under nitrogen, forming capsule-like beads where *in-situ* polymerization occurred. These beads exhibited enhanced functional group density, improved mechanical strength, and high adsorption efficiency for both single and mixed pollutants, enabling easy separation and reuse. A nanocomposite hydrogel was synthesized by grafting methacrylic acid (MAA) and acrylamide (AAm) onto a sodium alginate (SA) backbone via free radical polymerization (Safarzadeh et al., 2023). To boost dye adsorption and swelling capacity, montmorillonite (MMT) nanoparticles were added at 0–15 wt.%. The optimized composite (10 wt.% MMT) showed significantly enhanced malachite green dye removal and swelling behavior. Compared to physical methods, this chemical modification provided better control over the hydrogel's sorption properties.

5.2.1.3 Advanced Manufacturing using 3D Printing

3D printing enables precise, layer-by-layer fabrication of alginate-based structures with customized porosity and composition. Unlike conventional methods, it produces intricate architectures that improve surface area, mechanical strength, and sorption efficiency for targeted environmental applications. Conventional powdered adsorbents pose handling, recovery, and reuse challenges in large-scale applications. They often lead to high pressure drops and poor mass transfer in packed columns, with suboptimal packing configurations. In contrast, 3D printing enables the creation of structured sorbents with tailored porosity and enhanced mechanical strength, allowing for easier separation, reuse, and more efficient industrial deployment. A major benefit of additive manufacturing is the ability to create customized alginate-based biocomposites by precisely incorporating functional materials into bioinks. Design parameters like nozzle size, layer thickness, and infill density can be tailored to achieve desired sorbent properties. This clean, scalable method minimizes material waste—ideal for mass production. Consistent fabrication across batches enhances reproducibility and reliability in real-world applications, supporting field operators and engineers in deployment. Recent studies have explored functionalized 3D-printed alginate-based biocomposites for practical use. Magnetic variants embedded with iron oxide nanoparticles enable easy recovery and reuse (Kloster et al., 2020). Multi-material printing supports the creation of complex structures with targeted adsorption, ideal for treating challenging wastewater. Techniques like direct ink writing (DIW) offer a low-cost, versatile alternative to conventional, labor-intensive fabrication methods. Although 3D printed alginate-based adsorbents show strong potential, their practical application faces key challenges. Long-term stability in real aqueous conditions, resistance to leaching of embedded materials, and durability under continuous flow still require validation. Enhancements such as integrating pH-responsive or pollutant-indicating components could expand functionality—e.g., swelling in specific pH ranges or visible cues triggered by pollutant thresholds. However, these upgrades must preserve critical properties like rheology and printability to ensure successful fabrication and deployment. A recent study demonstrated that 3D-printed adsorbents made from calcium alginate and gelatin significantly improved methylene blue removal, especially when functionalized adsorbents were embedded (Pei et al., 2020). Scaffold geometry (e.g., square, hexagonal, circular) affected uptake by altering surface area and flow dynamics. While the base scaffold achieved ~43.6% removal, the functionalized variant reached 99.8% within 20 minutes due to synergistic microporosity and macrostructure. Unlike powdered adsorbents, which lose efficiency after two reuse cycles, the 3D-printed composites retained performance over seven cycles post acid regeneration.

A study developed 3D-printed sodium alginate/chitosan/halloysite nanotube composites with enhanced adsorption capabilities (Luo et al., 2024). Characterization confirmed a microporous structure and high surface area, aiding methylene blue (MB) removal. Under optimal conditions (1 g/L sorbent, 80 mg/L MB), 80% removal was achieved in 60 minutes. The maximum adsorption capacity reached 376.3 mg/g, highlighting the composite's strong affinity for dye molecules. A customized 3D-printable hydrogel ink incorporating alginate, gelatin, and polyethyleneimine (PEI) was developed for heavy metal removal (Finny et al., 2022). PEI introduced amine groups, boosting metal ion binding, while the hydrogel's rheological properties ensured mechanical stability in tablet form (Fig. 5.2). The porous structure provided high surface area and abundant binding sites, achieving 60% Cu^{2+} removal from 100 ppm in 5 h, rising to 98% at 18 h. At 1000 ppm, ~25% removal was observed. Adsorption followed the trend: $Cu^{2+} > Cd^{2+} > Co^{2+} > Ni^{2+}$. This rapid, scalable method yields cost-effective sorbents suitable for practical water treatment.

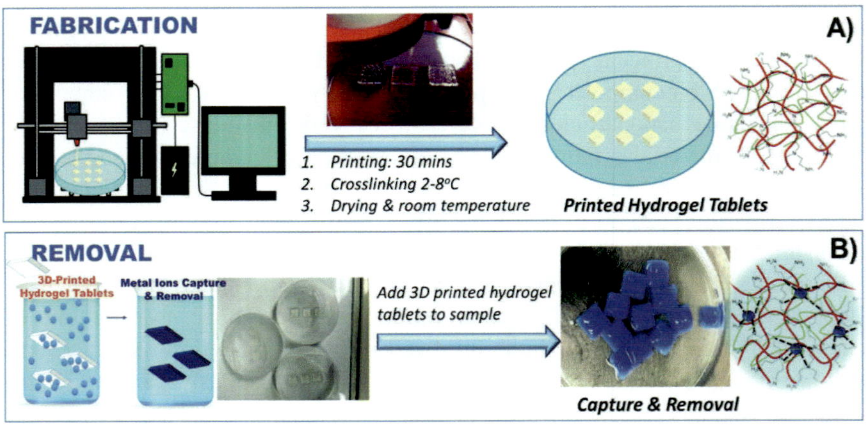

Figure 5.2: (A) Schematic of single-step 3D printing of hydrogel tablets; (B) removal mechanism showing interaction between hydrogel functional groups and metal ions. A visible color change indicates metal–ligand complex formation within the printed matrix (adapted from (Finny et al., 2022) under CC-BY license).

5.3 Characterization

Characterizing alginate-based biocomposites is key to understanding their structure–function relationships. Morphological techniques such as SEM, TEM, and AFM provide insights into surface texture, porosity, nanoparticle dispersion, and surface roughness. For magnetically modified composites, Vibrating Sample Magnetometry (VSM) assesses magnetic strength and coercivity. UV-Vis spectroscopy is also employed to evaluate optical absorption and transmission, supporting functional performance analysis across applications. Magnetic nanoparticle (MNP)-embedded hydrogels offer the benefit of easy recovery using external magnetic fields. However, conventional magnetic hydrogels often lack mechanical stability and become brittle under compact conditions like fixed-bed systems. To address this, a novel hollow-core hydrogel bead composite was developed, showing improved adsorption efficiency and enhanced mechanical durability (Salahuddin et al., 2022).

MNP-decorated cellulose nanofibers (CNF) were incorporated into an alginate-based matrix to create CNF/alginate and MNP–CNF/alginate hydrogel beads. The MNP–CNF variant formed a highly porous structure and exhibited superior mechanical strength, withstanding over 82% strain and >3.5% stress, unlike conventional alginate beads. Stress–strain tests confirmed their reusability in packed columns. Nitrogen isotherms revealed enhanced porosity and mesopore formation (30–100 nm) in MNP–CNF beads, supporting better nitrogen uptake. These features make them well-suited for continuous water treatment, guiding material optimization for effective remediation.

5.4 Applications

Alginate-based composites serve as versatile adsorbents for removing inorganic, organic and biological pollutants from water systems.

5.4.1 Removal of Inorganic Contaminants

The following subsections highlight representative studies on the use of alginate-based biocomposites for removing inorganic contaminants from water.

5.4.1.1 Heavy Metals and Metalloids

A recent study developed a sodium alginate (SA)-based nanofiber sorbent by incorporating polyvinylpyrrolidone (PVP) and Schiff base-functionalized $ZnO@SiO_2$ nanoparticles via electrospinning (Moridi and Gh, 2022). The Schiff base-modified $ZnO@SiO_2$, synthesized through a sol-gel process, was uniformly dispersed within the SA/PVP matrix, yielding fibers with an average diameter of 97 nm. Microscopy confirmed good nanoparticle distribution, while FTIR analysis indicated the presence of hydroxyl, amine, and Schiff base groups, contributing to enhanced metal ion binding. The nanofiber exhibited improved mechanical strength, aqueous stability, and high adsorption capacities for Cu(II) (289.5 mg/g) and Ni(II) (101.9 mg/g) at pH 6.0, 45°C, and 240 min. In a recent study, La^{3+} ions were embedded into sodium alginate fibers using wet spinning and ion exchange, followed by carbonization to produce La_2O_3-modified biochar for chromate remova (Sun et al., 2021). The material showed strong Cr(VI) adsorption, with a maximum capacity of ~104.9 mg/g at room temperature. Langmuir and pseudo-second-order models indicated monolayer chemisorption, while thermodynamic data confirmed a spontaneous, endothermic process. Removal mechanisms included electrostatic interaction, ligand exchange, and complexation. Biochar provided structural support and high surface area, enhancing La_2O_3 stability in low pH environments.

A 3D reduced graphene oxide/sodium alginate double-network (GAD) bead system was developed for Sb(III) removal from water (Yang et al., 2021). Graphene enhanced the composite's thermal stability and surface area, improving adsorption. Optimal performance (7.67 mg/g at 298 K) was observed under pseudo-second-order kinetics ($R^2 > 0.99$), with equilibrium aligning with Freundlich and Langmuir models. Adsorption decreased with rising temperature, indicating an exothermic process. The material maintained ~80% efficiency across pH 3–9 but showed reduced performance at high NaCl levels due to ion competition. Fixed-bed studies fit the Thomas model ($R^2 > 0.99$), and the beads remained effective over five reuse cycles.

5.4.1.2 Phosphate

Magnetic amorphous lanthanum silicate alginate hydrogel beads (MALS-B) were synthesized for phosphate removal from water (Chen et al., 2022). Mesoporous silica, derived from naturally sourced rectorite, was combined with lanthanum via a one-pot method to form amorphous lanthanum silicate (ALS). MALS-B were then fabricated by blending sodium alginate with ALS and Fe_3O_4 nanoparticles, followed by $CaCl_2$ cross-linking (Fig. 5.3A). For comparison, ALS-B beads were prepared identically without Fe_3O_4 incorporation.

ALS provided phosphate binding functionality, while Fe_3O_4 imparted magnetism to MALS-B beads for easy separation post-adsorption. Optical and SEM analyses (Fig. 5.3B) showed both ALS-B and MALS-B had uniform spherical shapes (~3 mm) and rugged surfaces with well-distributed particles. Structural integrity remained intact after freeze-drying. MALS-B exhibited a saturation magnetization of 17.38 emu/g and achieved high phosphate uptake (40.14 mg P/g), outperforming many hydrogel adsorbents. The beads also maintained strong selectivity and adsorption efficiency across a broad pH range.

5.4.1.3 Nitrate

Nitrate-contaminated water poses environmental risks like eutrophication, requiring efficient treatment. A study utilized hydrogel–rice husk biochar composites as low-cost, eco-friendly sorbents for nitrate removal (Afjeh et al., 2022). Composites with varying biochar contents (2.5–9.6%) were tested under different pH, time, and concentration conditions. Optimal performance—34.3% removal—was achieved at pH 3 using 0.02 g of the 5% biochar composite in 25 mL of 20 mg/L nitrate solution at 25°C over 60 minutes, balancing efficiency and economic viability. A study investigated nitrate removal using iron-modified

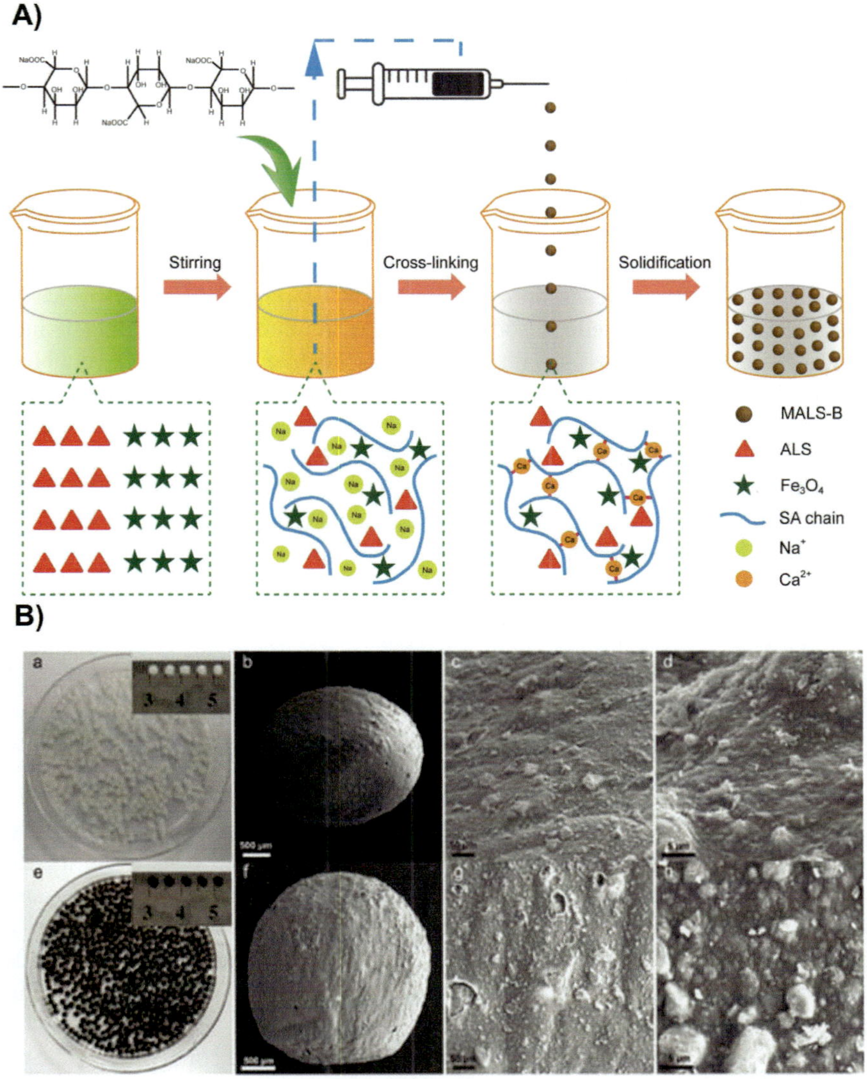

Figure 5.3: (A) Schematic illustration of the MALS-B synthesis process, (B) optical and SEM images of (a–d) ALS-B and (e–h) MALS-B beads. SEM scale bars: b and e = 500 µm; c and f = 50 µm; d and h = 5 µm (reproduced from (Chen et al., 2022) under CC-BY license).

clinoptilolite beads with alginate (Fe/Clin-A) and chitosan (Fe/Clin-Ch) matrices (Hosseini and Mousavi, 2023). Fe/Clin-Ch beads achieved 88.79% removal at 3 g/L in 60 min, outperforming Fe/Clin-A beads, which reached 86.13% at 4 g/L in 75 min. Langmuir model analysis showed higher adsorption capacities for Fe/Clin-Ch (36.63 mg/g) than Fe/Clin-A (24.39 mg/g). According to the

Dubinin–Radushkevich model, respective capacities were 10.93 mg/g and 7.06 mg/g.

Immobilizing nanoparticles in alginate-based hydrogels enhances their reactivity and ease of recovery in wastewater treatment. Calcium-alginate beads effectively encapsulate nanoparticles but tend to sink due to high density, complicating retrieval. To address this, CO_2 bubbles are generated and trapped during synthesis, reducing bead density and enabling flotation. This simple method greatly improves bead recovery, while alginate concentration variations have minimal effect on buoyancy.

Calcium-alginate beads were fabricated by mixing PAC, NZVI, MNP, and calcium carbonate into a sodium alginate solution, then dropping it into a $CaCl_2$–acetic acid bath to form hydrogels with embedded CO_2 bubbles, reducing bead density (Bahrami et al., 2020). SEM and optical imaging confirmed uniform nanoparticle distribution and bubble incorporation. These beads removed 4.3–9.6 mg NO_3-N via 1 g immobilized NZVI within 48 h, while PAC adsorbed up to 47.8% of generated NH_4-N. This dual-functional system offered efficient nitrate remediation and potential ammonia recovery for use as fertilizer, supporting both water treatment and nutrient recycling.

5.4.2 Removal of Organic Contaminants

Biochar-based composites have shown strong efficacy in treating wastewater containing various organic pollutants, as detailed in the following sections.

5.4.2.1 Dyes

A lignin–sodium alginate hydrogel adsorbent was synthesized via Ca^{2+} cross-linking to enhance methylene blue (MB) removal (Chen et al., 2022). The composite showed improved thermal stability and achieved peak performance at pH 12 and 45°C, with a maximum adsorption capacity of 254.3 mg/g and 84.8% removal efficiency. Thermodynamic analysis confirmed an endothermic process, while kinetics followed a chemisorption-driven model. Notably, the beads maintained > 90% MB removal over five regeneration cycles, demonstrating strong reusability.

Nanocomposite films were synthesized via UV-induced copolymerization of methacrylated alginate (MALG) and acrylic acid (AA), reinforced with graphene oxide (GO) for improved dye removal (Teijido et al., 2024). GO enhanced mechanical stability, raising storage modulus from 676.6–1538.7 MPa (25°C) to 886.9–2066.6 MPa (100°C). The films showed rapid water absorption, ~30 min drying time, and high hydrophilicity. MB adsorption efficiency reached 70–80% in 30 h, increasing to 85–98% after 258 h. Reusability tests under acidic conditions confirmed sustained performance over multiple cycles. A study synthesized SA-g-poly(MAA-co-AAm) hydrogels with montmorillonite (MMT) nanoparticles

(0–15 wt.%) to enhance malachite green (MG) adsorption (Safarzadeh et al., 2023). Optimal performance was achieved at 10 wt.% MMT, showing a maximum capacity of 95.91 mg/g at pH 7, 1.5 g/L sorbent dose, 298 K, and 90 min contact time for 10 mg/L MG. The process followed Langmuir isotherm and pseudo-second-order kinetics. Thermodynamic analysis confirmed spontaneity and exothermicity, with $\Delta H°$ values of –61.5 kJ/mol (unmodified) and –77.3 kJ/mol (MMT-modified).

5.4.2.2 Pesticides

A chitosan–alginate adsorbent synthesized via microemulsion was tested for malathion removal from water (Sabbagh et al., 2021).Optimal conditions—0.18 g adsorbent, 10 mg/L malathion, and 20 min contact—yielded 82.35% removal. Response surface methodology confirmed model accuracy. The process followed Langmuir isotherm ($R^2 = 0.99$), indicating monolayer adsorption, and pseudo-second-order kinetics ($R^2 = 0.99$), suggesting chemisorption as the dominant mechanism. A study utilized alginate-encapsulated biochar derived from sugarcane bagasse for chlorpyrifos removal from water, demonstrating a green and low-cost remediation approach (Jacob et al., 2022). Optimal performance—86% removal and 6.25 mg/g adsorption capacity—was achieved at pH 10 with 10 mg/L chlorpyrifos. Adsorption followed the Freundlich isotherm ($R^2 = 0.9332$), indicating multilayer uptake, and intraparticle diffusion kinetics ($R^2 = 0.9750$), with the Boyd model confirming diffusion as the rate-limiting step. Thermodynamic analysis revealed an exothermic, physisorption-driven process.

5.4.2.3 Pharmaceutical Products

An integrated adsorbent–photocatalyst system using calcium alginate beads embedded with S-doped TiO_2 and WS_2 (S-TiO_2/WS_2) was developed for oxytetracycline (OTC) removal (Kumar et al., 2022). Synthesized via hydrothermal and ionic gelation methods, the beads achieved 92.5% removal at 10 mg/L OTC, decreasing to 72% at 75 mg/L. The synergy between adsorption and photocatalysis enabled efficient OTC degradation via reactive radicals. Scavenger tests confirmed the key role of holes and hydroxyl radicals, while $K_2S_2O_8$ boosted degradation via additional oxidants. The beads maintained 78.6% efficiency after five reuse cycles, highlighting their stability and scalability. A 3D-printed alginate hydrogel incorporating 10 wt.% rice husk biochar was developed to enhance sorption of ibuprofen and methylene blue in tap water (Silva et al., 2023). Characterization confirmed increased porosity and swelling capacity. The composite achieved adsorption capacities of 111.4 mg/g for ibuprofen and 214.6 mg/g for methylene blue—representing 48% and 58% improvements over plain hydrogels. Sorption followed pseudo-first-order kinetics and the Freundlich isotherm. The material retained performance across 20 reuse cycles, demonstrating strong reusability and stability.

5.4.3 Removal of Biological Contaminants

A Zn^{2+}-crosslinked alginate/nanohydroxyapatite (ZA/nHA) composite, derived from *S. latifolium* alginate and *T. fragilis*-based nHA, was developed for biological contaminant removal (Gomaa and Danial, 2023). At 0.5–1% (w/v), it effectively disinfected *E. coli* ($\leq 10^5$ CFU/mL) within 300–360 min, with disinfection kinetics aligning with the Weibull model. The composite showed strong anti-algal activity, removing 100% of Chroococcus sp. (2 days) and 90% of Chlorella sp. (4 days). It also rapidly adsorbed crystal violet dye (0.171 mg/g in 7 min) at neutral pH (6–7), making it practical for broad-spectrum water treatment.

A study encapsulated nanosilica derived from rice husk into sodium alginate beads for antibacterial water treatment (Ravindran et al., 2022).Characterization confirmed nanosilica incorporation, and disinfection was tested via Petrifilm and column methods. Beads with 1.25 g nanosilica achieved 92% bacterial reduction within 20 min retention time, showing strong efficacy against E. coli. The low-cost, bio-based beads present a viable alternative to conventional nanomaterial disinfectants. A study developed alginate beads encapsulating ZnO nanoflowers and activated carbon (AC) for removing antibiotic-resistant bacteria (ARB) and resistance genes (ARGs) from wastewater (Liu et al., 2023). Fixed-bed columns with 1 g ZnO (~2 g/L) showed strong removal efficiency and low Zn^{2+} leaching. Operating at 1 mL/min with 10^7 CFU/mL influent, the system achieved a 5.69-log ARB and 2.44–2.74-log ARG reduction. Mechanistic analysis linked ZnO's bactericidal action to ROS generation, membrane disruption, and cytoplasmic leakage.

5.5 Challenges and Prospects

Alginate-based biocomposites show great promise in environmental remediation, but practical challenges remain. Their tendency to swell and degrade in water limits long-term use, while leaching of embedded materials can lower adsorption efficiency and raise ecotoxicity concerns, especially with nanomaterials (Eltaweil et al., 2022). Poor immobilization and degradation in acidic or high-ionic conditions may release harmful species like Zn^{2+} from ZnO nanoflowers. Leached metals and carbon-based residues could disrupt ecosystems or health. Ensuring structural durability, process-specific selectivity, and synthesis reproducibility is vital. Moreover, regeneration methods and techno-economic feasibility need thorough evaluation before large-scale deployment. These challenges highlight the need for advancing alginate-based biocomposites. Incorporating magnetic components can aid post-treatment separation, while biodegradable coatings or stabilizers may reduce leaching and enhance eco-compatibility. Though batch tests offer initial insights, column studies under real-world conditions are essential for evaluating long-term performance. Integrating adsorption with complementary processes like photocatalysis could enable modular, decentralized water treatment. Future development should include life cycle assessments and emphasize zero liquid discharge principles.

5.6 Conclusion

Alginate-based biocomposites offer a sustainable approach to pollutant removal, with functional additives enhancing their water purification performance. Advances in 3D printing enable tailored, durable, and reusable adsorbents for both batch and continuous systems. Their biodegradable nature supports low-waste manufacturing. While issues like durability, leaching, and scalability remain, progress in functionalization, magnetic recovery, and fixed-bed integration continues to improve practical viability for sustainable water treatment.

References

Anjana, K. and Arunkumar, K. 2024. Brown algae biomass for fucoxanthin, fucoidan and alginate; update review on structure, biosynthesis, biological activities and extraction valorisation. *Int. J. Biol. Macromol.*, 280: 135632.

Awasthi, A., Maheshwari, V., Rastogi, K., Singh, K. Kumar, S. et al. 2023. Treatment of textile dyes using modified banyan leaf-based biosorbent using batch and intensified batch process: Kinetic and isotherm studies. *Biomass Convers. Biorefin.*, 1(2023): 1–16.

Bahrami, F., Yu, X., Zou, Y., Sun, Y., Sun, G. et al. 2020. Impregnated calcium-alginate beads as floating reactors for the remediation of nitrate-contaminated groundwater. *Chemical Engineering Journal*, 382: 122774.

Chen, H., Zeng, H. and Yang, H. 2022. Phosphate removal from wastewater by magnetic amorphous lanthanum silicate alginate hydrogel beads. *Minerals*, 12: 171.

Chen, T., Liu, H., Gao, J., Hu, G., Zhao, Y. et al. 2022. Efficient removal of methylene blue by bio-based sodium alginate/lignin composite hydrogel beads. *Polymers*, 14(2022): 2917.

Cho, D.W., Song, H., Kim, B., Schwartz, F.W., Jeon, B.H. et al. 2015. Reduction of nitrate in groundwater by Fe(0)/Magnetite nanoparticles entrapped in Ca-Alginate beads. *Water Air Soil Pollut.*, 226: 1–10.

Dechojarassri, D., Omote, S., Nishida, K., Omura, T., Yamaguchi, H. et al. 2018. Preparation of alginate fibers coagulated by calcium chloride or sulfuric acid: Application to the adsorption of Sr2+. *J. Hazard. Mater.*, 355: 154–161.

Doyo, A.N., Kumar, R. and Barakat, M.A. 2023. Recent advances in cellulose, chitosan, and alginate based biopolymeric composites for adsorption of heavy metals from wastewater. *J. Taiwan Inst. Chem. Eng.*, 151: 105095.

El-Aassar, M.R., Al-derbas, N.A., Al-Muaikel, N.S., Rafea, M.A., Alhumaimess, M.S. et al. 2024. Development and characterization of sodium alginate beads incorporating functionalized poly(acrylonitrile-co-styrene)/carbon nanotubes for effective Fe(II) ion adsorption from aqueous solutions. *Int. J. Biol. Macromol.*, 283: 137947.

Eltaweil, A.S., Abd El-Monaem, E.M., Elshishini, H.M., El-Aqapa, H.G., Hosny, M. et al. 2022. Recent developments in alginate-based adsorbents for removing phosphate ions from wastewater: A review. *RSC Adv.*, 12: 8228–8248.

Faouzia, B., Imene, K., Chaima, H., Riadh, B., Imene, B.A., Gil, A. et al. 2025. Sustainable biomass-derived activated biochar/polyaniline/alginate composite beads for enhanced methylene blue adsorption: Experimental and theoretical investigations. *Int. J. Biol. Macromol.*, 305: 141338.

Fernando, I.P.S., Lee, W.W., Han, E.J. and Ahn, G. 2020. Alginate-based nanomaterials: Fabrication techniques, properties, and applications. *Chemical Engineering Journal*, 391: 123823.

Finny, A.S., Cheng, N., Popoola, O. and Andreescu, S. 2022. 3D printable polyethyleneimine based hydrogel adsorbents for heavy metal ions removal. *Environmental Science: Advances*, 1: 443–455.

Gomaa, M. and Danial, A.W. 2023. Seaweed-based alginate/hydroxyapatite composite for the effective removal of bacteria, cyanobacteria, algae, and crystal violet from water. *J. Biol. Eng.*, 17: 1–14.

Guo, H., Qin, Q., Chang, J.S. and Lee, D.J. 2023. Modified alginate materials for wastewater treatment: Application prospects. *Bioresour. Technol.*, 387: 129639.

Hastuti, F.W., Pradyasti, A., Lee, M.J., Hyun, W.J., Kim, H.J. et al. 2025. Continuous flow catalytic water remediation using Pd and Pd-Pt nanoplates-functionalized alginate/PVA hydrogels. *Journal of Water Process Engineering*, 71: 107427.

Hosseini Nami, S. and Mousavi, S.B. 2023. Nitrate removal performance of different granular adsorbents using a novel fe-exchanged nanoporous clinoptilolite. *Ind. Eng. Chem. Res.*, 62: 3659–3671.

Jacob, M.M., Ponnuchamy, M., Kapoor, A. and Sivaraman, P. 2022. Adsorptive decontamination of organophosphate pesticide chlorpyrifos from aqueous systems using bagasse-derived biochar alginate beads: Thermodynamic, equilibrium, and kinetic studies. *Chemical Engineering Research and Design*, 186: 241–251.

Kausar, A., Sher, F., Hazafa, A., Javed, A., Sillanpää, M. et al. 2020. Biocomposite of sodium-alginate with acidified clay for wastewater treatment: Kinetic, equilibrium and thermodynamic studies. *Int. J. Biol. Macromol.*, 161: 1272–1285.

Khoj, M.A., Hassan, A.F., Awwad, N.S., Ibrahium, H.A., Shaltout, W.A. et al. 2024. Fabrication and characterization of Araucaria gum/calcium alginate composite beads for batch and column adsorption of lead ions. *Int. J. Biol. Macromol.*, 255: 128234.

Kloster, G.A., Muraca, D., Moscoso Londoño, O., Pirota, K.R., Mosiewicki, M.A. et al. 2020. Alginate based nanocomposites with magnetic properties. *Compos. Part A Appl. Sci. Manuf.*, 135: 105936.

Kumar, R., Ansari, M.O., Taleb, M.A., Oves, M., Barakat, M.A. et al. 2022. Integrated adsorption-photocatalytic decontamination of oxytetracycline from wastewater using S-doped TiO_2/WS_2/ calcium alginate beads. *Catalysts*, 12: 1676.

Liu, Z., Yu, X., Zhou, Z., Zhou, J., Shuai, X. et al. 2023. 3D ZnO/activated carbon alginate beads for the removal of antibiotic-resistant bacteria and antibiotic resistance genes. *Polymers (Basel.)*, 15: 2215.

Luo, J., Ji, A., Xia, G., Liu, L. and Yan, J. 2024. Construction of 3D-printed sodium alginate/chitosan/ halloysite nanotube composites as adsorbents of methylene blue. *Molecules*, 29: 1609.

Luo, J., Ji, A., Xia, G., Liu, L., Yan, J. et al. 2024. Construction of 3D-printed sodium alginate/chitosan/ halloysite nanotube composites as adsorbents of methylene blue. *Molecules*, 29(2024): 1609.

Mohammed, C., Mahabir, S., Mohammed, K., John, N., Lee, K.Y. et al. 2019. Calcium alginate thin films derived from sargassum natans for the selective adsorption of Cd2+ , Cu2+ , and Pb2+ ions. *Ind. Eng. Chem. Res.*, 58: 1417–1425.

Mojiri, A. and Bashir, M.J.K. 2022. Wastewater treatment: Current and future techniques. *Water*, 14: 448.

Moridi, H. and Gh, A.B. 2022. Sodium alginate/polyvinyl pyrrolidone/zinc oxide @silica Schiff-base nanofiber membrane for single and binary removal of copper and nickel cations from aqueous medium. *Research on Chemical Intermediates*, 48: 4643–4670.

Muthukumaran, P., Suresh Babu, P., Karthikeyan, S., Kamaraj, M., Aravind, J. et al. 2021. Tailored natural polymers: A useful eco-friendly sustainable tool for the mitigation of emerging pollutants: A review. *International Journal of Environmental Science and Technology*, 18(2021): 2491–2510.

Pei, R., Fan, L., Zhao, F., Xiao, J., Yang, Y. et al. 2020. 3D-Printed metal-organic frameworks within biocompatible polymers as excellent adsorbents for organic dyes removal. *J. Hazard Mater.*, 384: 121418.

Pinheiro, C.P., Moreira, L.M.K., Alves, S.S., Cadaval, T.R.S., Pinto, L.A.A. et al. 2021. Anthocyanins concentration by adsorption onto chitosan and alginate beads: Isotherms, kinetics and thermodynamics parameters. *Int. J. Biol. Macromol.*, 166: 934–939.

Rashid, R., Shafiq, I., Akhter, P., Iqbal, M.J., Hussain, M. et al. 2021. A state-of-the-art review on wastewater treatment techniques: The effectiveness of adsorption method. *Environmental Science and Pollution Research*, 28(2021): 9050–9066.

Ravindran, L., Jesitha, K., Megha, P.U., Anilkumar, S., Sreekala, M.S. et al. 2022. Nanosilica entrapped alginate beads for the purification of groundwater contaminated with bacteria. *Silicon*, 14: 8707–8720.

Sabbagh, N., Tahvildari, K. and Mehrdad Sharif, A.A. 2021. Application of chitosan-alginate bio composite for adsorption of malathion from wastewater: Characterization and response surface methodology. *J. Contam. Hydrol.*, 242: 103868.

Sadeghi Afjeh, M., Bagheri Marandi, G. and Zohuriaan-Mehr, M.J. 2020. Nitrate removal from aqueous solutions by adsorption onto hydrogel-rice husk biochar composite. *Water Environment Research*, 92: 934–947.

Safarzadeh, H., Peighambardoust, S.J. and Peighambardoust, S.H. 2023. Application of a novel sodium alginate-graft-poly(methacrylic acid-co-acrylamide)/montmorillonite nanocomposite hydrogel for removal of malachite green from wastewater. *Journal of Polymer Research*, 30: 1–14.

Salahuddin, B., Aziz, S., Gao, S., Hossain, M.S.A., Billah, M. et al. 2022. Magnetic hydrogel composite for wastewater treatment. *Polymers*, 14(2022): 5074.

Silva, E.C., Soares, V.R., Nörnberg, A.B. and Fajardo, A.R. 2023.Recyclable 3D-printed composite hydrogel containing rice husk biochar for organic contaminants adsorption in tap water. *ACS Appl. Polym. Mater.*, 5: 8415–8429.

Sun, X., Guo, P., Sun, Y. and Cui, Y. 2021. Adsorption of hexavalent chromium by sodium alginate fiber biochar loaded with lanthanum. *Materials*, 14: 2224.

Teijido, R., Zhang, Q., Blanco, M., Pérez-Álvarez, L., Lanceros-Méndez, S. et al. 2024. Graphene-enhanced methacrylated alginate gel films for sustainable dye removal in water purification. *Gels*, 10: 25.

Wang, X., Zhang, H., He, Q., Xing, H., Feng, K. et al. 2022. Core-shell alginate beads as green reactor to synthesize grafted composite beads to efficiently boost single/co-adsorption of dyes and Pb(II). *Int. J. Biol. Macromol.*, 206: 10–20.

Xin, L., Jia, A., Liu, Z., Shi, C., Liu, Z. et al. 2025. Covalent organic polymer-based sodium alginate composite beads: Decontamination of ciprofloxacin micro-polluted groundwater in a PRB system. *Sep. Purif. Technol.*, 354: 128988.

Yang, X., Zhou, T., Deng, R., Zhu, Z., Saleem, A. et al. 2021. Removal of Sb(III) by 3D-reduced graphene oxide/sodium alginate double-network composites from an aqueous batch and fixed-bed system. *Scientific Reports*, 11(2021): 1–17.

Zdiri, K., Cayla, A., Elamri, A., Erard, A. and Salaun, F. 2022. Alginate-based bio-composites and their potential applications. *Journal of Functional Biomaterials*, 13: 117.

CHAPTER

6 Mechanical and Wear Property Enhancement of GFRP Composite through Filler Addition

Manoj Kumar[1] and *Hemalata Jena*[2,*]

Glass Fibre Reinforced Polymer (GFRP) has gained significant attention or its advantages over metals, with applications ranging from household items to aerospace. Introducing fillers as a third phase in the polymer matrix enhances GFRP's mechanical and tribological properties, especially under dynamic and harsh conditions. This study systematically reviews how various fillers influence GFRP performance, highlighting their potential for advanced engineering, industrial, and marine applications.

6.1 Introduction

Fiber Reinforced Polymers (FRPs), especially Glass Fiber Reinforced Polymers (GFRPs), are increasingly replacing metals in the automotive sector due to their superior stiffness-to-weight ratio, cost-effectiveness, and fuel efficiency. Reinforcement with synthetic fibers such as E-glass, carbon, Kevlar, and aramid significantly enhances mechanical performance (Shivanagere et al., 2018; Mohammadi et al., 2022; Das et al., 2014). GFRPs find broad application across industries—from aerospace and marine to biomedicine and oil & gas—

[1] Department of Mechanical Engineering, RTC Institute of Technology, Ormanjhi, India.
[2] School of Mechanical Engineering, KIIT Deemed to be University, Bhubaneswar, India.
* Corresponding author: hemalata.jenafme@kiit.ac.in

DOI: 10.1201/9781003607359-6

owing to their high strength, corrosion resistance, elasticity, and thermal stability. Composed primarily of matrix and fibers, these composites can be further advanced using hybrid reinforcements that combine multiple fiber or particle types. This hybridization of composite materials can be accomplished by using dissimilar types of fibers as different layers, including dissimilar types of fibers by interment layers of two different fibers, or using dissimilar matrix materials in different regions of the same structure (Thakur et al., 2017; Turaka and Bandaru, 2023. Each constituent is contributing their unique properties to create a composite with enhanced overall performance (Hulugappa et al., 2016; Espert et al., 2004; Palanikumar et al., 2012). Developing hybrid composites involves complexities such as ensuring material compatibility, interfacial adhesion, and suitable processing methods. Challenges like thermal expansion mismatch and galvanic corrosion also arise. Despite this, hybrid composites can outperform both single-phase and conventional composites. Kumar et al. (2025) demonstrated that a gradient-toughened system improved ductility and erosion resistance, emphasizing the need for tailored material modifications to meet specific performance criteria. Filler incorporation has proven effective in addressing key limitations of GFRP composites, opening new avenues in composite engineering. Studies highlight the importance of optimizing filler content through experimental validation, modeling, and data analysis (Nwobi et al., 2020; Wibawa et al., 2024). This approach enables designers to tailor hybrid composite configurations to achieve target properties such as strength, stiffness, and thermal or electrical conductivity for specific applications.

Enhanced Mechanical Properties: By combining reinforcing fibers or particulate fillers with varied mechanical characteristics—like strength, stiffness, and toughness—hybrid composites outperform single-component systems, offering improved strength-to-weight ratios and enhanced impact resistance (Wibawa, 2024; Mohanty, 2024).

Cost Efficiency: Using cost-effective strategies, combining high-performance fibers like carbon with lower-cost alternatives can reduce raw material expenses while maintaining desirable performance levels (Del Vecchio et al., 2024).

Improved Fatigue Resistance: Incorporating varied fibers, fillers, and matrices enhances the fatigue resistance of hybrid composites, improving their performance under cyclic loading—crucial for applications with repeated stress (Wang, 2020).

Damage Tolerance: Composite materials generally show strong damage tolerance and occasional resistance to delamination (Lazar, 2024). Additionally, the inhomogeneous nature of hybrid composites helps hinder crack propagation, reducing the risk of sudden failure.

Environmental Sustainability: Hybrid composites can incorporate sustainable or bio-based components, reducing reliance on non-renewable resources and supporting greener manufacturing practices (Ascione et al., 2024).

Multifunctionality: Hybrid composites enable the design of multifunctional materials, integrating structural strength with properties like electrical or thermal insulation—broadening their application potential (Ganeshkumar, 2024).

Versatility: Hybrid composites have broad applicability across aerospace, automotive, marine, sports, and civil engineering sectors, thanks to their adaptable mixed-material configurations (Das et al., 2019).

Weight Reduction: Careful selection of constituent materials in hybrid composites enables significant weight reduction compared to conventional materials—an essential advantage for enhancing fuel efficiency and overall performance in aerospace and automotive sectors (Morin, 2009).

In summary, hybrid composites combine the best attributes of diverse materials while minimizing their limitations, making them ideal for a wide range of high-performance applications. Their adaptability offers ongoing opportunities for innovation in materials, fabrication methods, and novel uses—pushing the boundaries of engineering design.

6.2 Applications of Hybrid Composite Materials

These materials have a diverse range of uses thanks to their improved performance attributes. Here are a few examples as shown in Fig. 6.1 of hybrid composite materials which can be applied:

Aerospace Industry: In aerospace, hybrid composites are extensively used in structural components like wings, panels, and fuselage sections due to their high

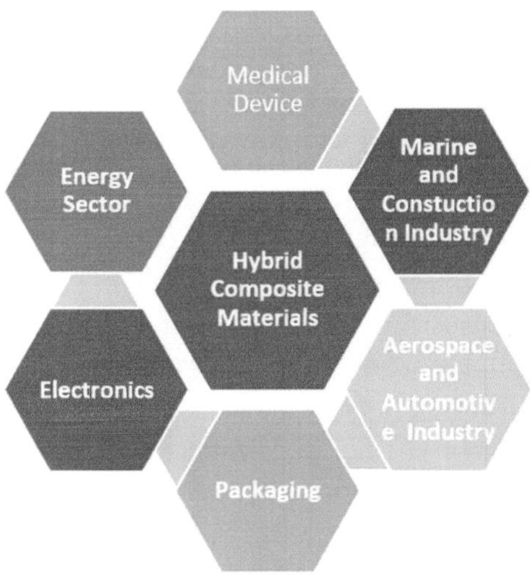

Figure 6.1: Application of hybrid composite.

strength-to-weight ratio—key to boosting fuel efficiency and flight performance (Puttegowda et al., 2018; Edwin Raja Dhas and Arun, 2022).

Automotive Sector: Hybrid composites are increasingly used in automotive components such as body panels, chassis parts, and interiors, contributing to significant weight reduction—ultimately enhancing fuel efficiency and lowering emissions (Ravishankar et al., 2019).

Marine Industry: Hybrid composites are widely used in catamaran hulls, decks, and marine structures due to their excellent corrosion resistance, high strength-to-weight ratio, and enhanced durability in harsh marine environments (Neşer, 2017).

Construction Industry: Hybrid composites are widely applied in construction for lightweight, high-strength structures, particularly in cladding, roofing, and seismic retrofitting (Knippers and Waimer, 2015). In civil engineering, they enhance bridges, columns, and beams by offering superior load-bearing capacity, long-term durability, and reduced maintenance needs (Madhavan et al., 2021).

Medical Devices: Hybrid composites are also used in medical fields such as prosthetics, orthotics, dental implants, and surgical tools. These materials can be engineered for specific biocompatibility, strength, and flexibility requirements (Ates et al., 2017; Zamora-Mendoza et al., 2023).

Energy Sector: Hybrid composites are used in wind turbine blades and lightweight solar panel structures to boost energy efficiency and support sustainable energy systems (Reddy et al., 2021; Trzepieciński et al., 2023). They're also applied in hydrokinetic turbine blades and geothermal wellbore components (León Gómez et al., 2023).

Packaging: Lightweight hybrid composites in packaging enhance protection, reduce weight, and improve environmental performance. They also extend product lifespan and help lower ecological impact (Majeed et al., 2013; Kenyó et al., 2018).

Electronics: Hybrid composites are used in electronic devices to produce lightweight, durable casings with added benefits like electromagnetic shielding and thermal regulation (Cetin and Karahan Toprakci, 2021; Petrény et al., 2033).

Hybrid composites offer remarkable versatility by combining two or more materials, enabling tailored properties for diverse sector-specific needs (Gu et al., 2018). Their use spans electronics casings, luggage, and furniture, where they blend strength, design, and lightweight performance (Seydibeyoğlu et al., 2023). In sports, they're found in tennis rackets, bicycles, and skis, delivering sturdiness, flexibility, and vibration damping (Ong et al., 2023). Defense applications include armor panels and combat helmets, where hybrid materials enhance ballistic and impact resistance (Zhang, 2015; Harussani et al., 2022).

6.3 GFRP Composites with Particulate Fillers

Glass Fiber-Reinforced Polymer (GFRP) composites offer a broad spectrum of mechanical properties, making them essential in aerospace, automotive, and other industries. Their strength and abrasion resistance stem from high-tensile glass fibers combined with lightweight polymer resins. Mechanical performance can be further improved by adding fillers like salinized silica nanoparticles, which enhance interfacial bonding. Surface treatments such as silane coating also boost fiber-matrix adhesion, increasing wear resistance and extending service life. As a result, GFRP-based composites are pivotal in meeting modern engineering demands across diverse applications. Hybrid glass fibre-reinforced composites combine glass fibres with various fillers to enhance properties like strength and wear resistance. Fibres can be natural (e.g., jute, flax) or synthetic (e.g., carbon, Kevlar). Adding fillers—an emerging method in the polymer industry—further boosts mechanical and tribological performance. These fillers range in scale (macro, micro, nano) and type: natural (e.g., cellulose) or synthetic (e.g., glass beads, ceramics). Nano-fillers such as carbon nanotubes and graphene, along with micro-fillers like microspheres, offer tailored improvements for advanced applications.

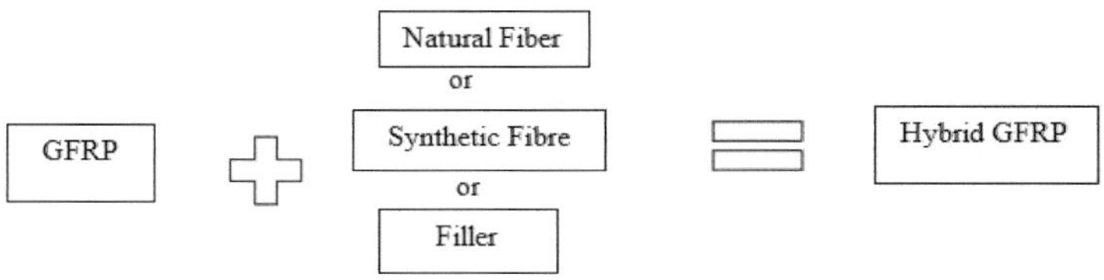

Figure 6.2: Formation of hybrid GFRP.

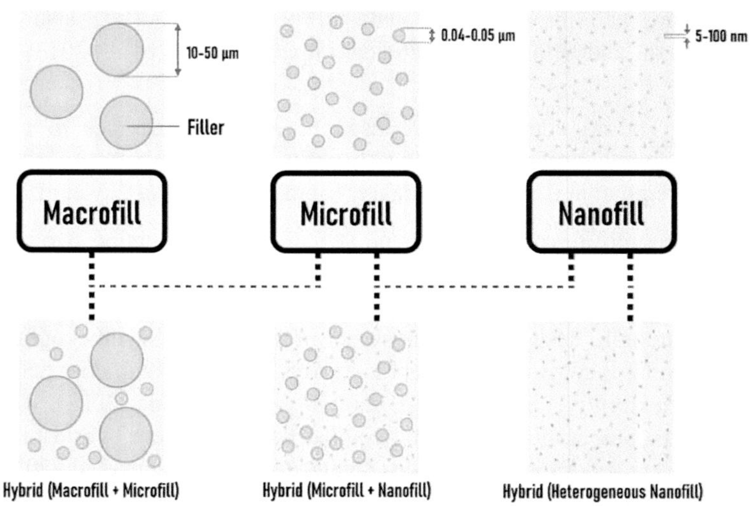

Figure 6.3: Fillers of various sizes in polymer (Czech et al., 2021) (Open Access).

The combination of different fillers can lead to synergistic effects, enhancing the overall performance of the composite. Figure 6.4 shows the addition of fillers in GFRP. Here are a few examples of hybrid glass fibre reinforced composites with different fillers:

Carbon Nanotubes (CNTs): Integrating carbon nanotubes (CNTs) into glass fiber composites enhances mechanical properties like stiffness and strength due to their high aspect ratio and superior reinforcement capability (Dehrooyeh et al., 2021).

Nanoclay: Incorporating nanoclay into glass fiber composites enhances flame retardancy, barrier properties, and dimensional stability. It also improves gas permeation resistance and thermal stability (Kowshik et al., 2022).

Silica Nanoparticles: Adding silica nanoparticles to glass fiber composites enhances wear resistance, thermal stability, and electrical conductivity. They also improve interfacial adhesion between fibers and the matrix (Karnati et al., 2020).

Graphene: Adding graphene to glass fiber composites enhances electrical conductivity, thermal behavior, and mechanical performance, even at low loading levels—thanks to its strong two-dimensional reinforcement structure (Singh and Kumar, 2021).

Metallic Particles: Adding metallic particles like aluminum or titanium to composites boosts electrical and thermal conductivity, while also improving mechanical properties such as fatigue resistance (Elsheikh and Yakoub, 2020).

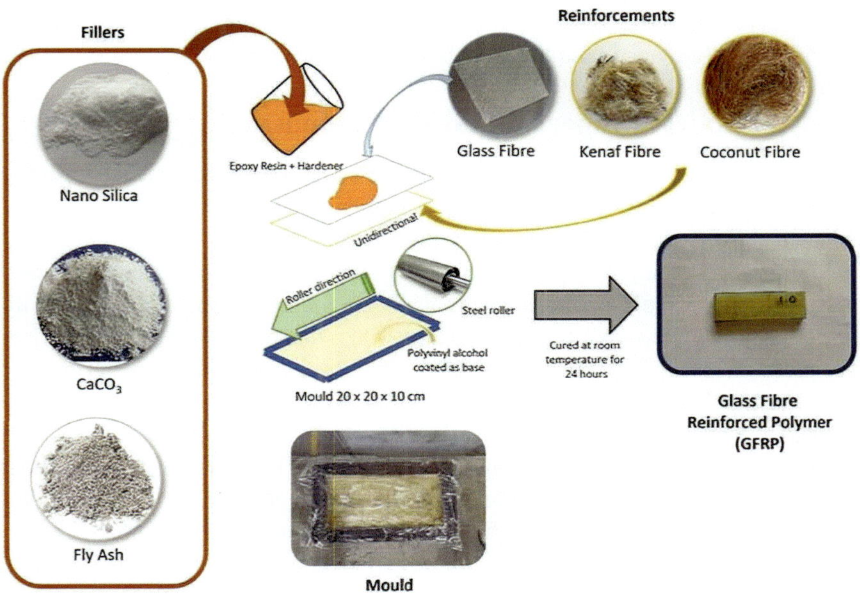

Figure 6.4: Hybridisation of GFRP through filler addition (Yadav et al., 2023; Badaruzzaman et al., 2022) (Open Access).

Ceramic Particles: The incorporation of ceramic particles like alumina (Al_2O_3) or silicon carbide can give higher resistance in wear condition and high temperature (Das et al., 2019). These particles also contribute to increased hardness and thermal conductivity (Agarwal et al., 2013).

Natural Fillers (Rice Husk Ash, Bagasse, etc.): Using agricultural waste fillers like rice husk ash or bagasse fibers enhances composite sustainability by reducing environmental impact and offering unique functional properties (Arjmandi et al., 2015; Suhot et al., 2021; Ayswarya et al., 2021).

Recycled Materials: Recycled rubber or plastic waste fillers in glass fiber composites enhance sustainability by reducing waste while adding properties like vibration damping and energy absorption. They also help disperse impact forces, lowering the risk of fracture and delamination (Genovés et al., 2022; Nagaraja et al., 2018).

Microglass Spheres: Microglass spheres—tiny solid glass particles—are used in composites to control density and mechanical properties. They also act as thermal insulators, enhancing composite strength (Agrawal et al., 2018; Yung et al., 2009).

Wollastonite Particles: Wollastonite, a natural calcium silicate mineral, has been shown to enhance the tensile, flexural, and impact properties of glass and sisal fiber composites (Karle and Tungikar, 2021; Singh et al., 2003).

Natural Minerals (Mica, Talc): Mica and talc, as natural mineral fillers, are effective in modifying the mechanical and thermal properties of glass fiber composites. Talc addition, in particular, has been shown to enhance overall composite performance (Kuram, 2019).

6.4 Mechanical Property of GFRP Composites

Numerous studies highlight how different fillers affect the mechanical properties of GFRP composites. Nano-fillers like graphene at 1–3 wt.% significantly enhance tensile strength due to their high aspect ratio and strong interfacial bonding (Seeli et al., 2023). Natural fillers such as wheat husk, coconut coir, and rice husk have also been used, with glass-epoxy composites generally outperforming glass-polyester ones—except when rice husk is used. Micro-fillers like glass microspheres improve flexural strength, contributing to better overall performance (Ferreira et al., 2010). Various fillers significantly influence GFRP composite performance. Graphene (1–3 wt.%) improves tensile strength through high aspect ratio and strong bonding (Seeli et al., 2023). Natural fillers like wheat husk and coconut coir enhance tensile properties, with glass-epoxy outperforming glass-polyester—except with rice husk. Micro-fillers such as glass microspheres boost flexural strength and overall durability (Ferreira et al., 2010).

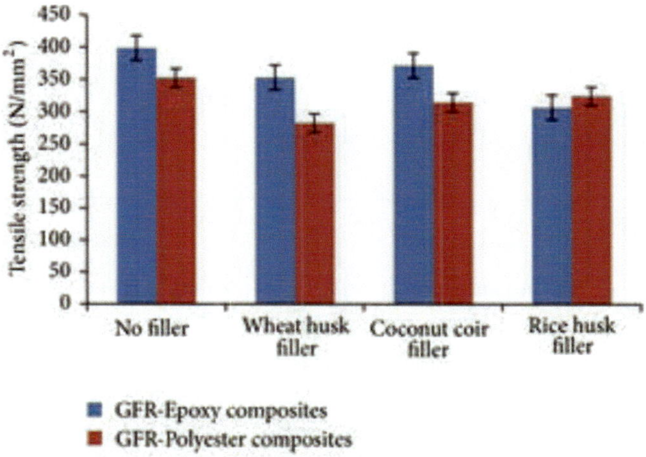

Figure 6.5: Composites' tensile strength with and without filler (Dhawan et al., 2013) (Open Access).

Mechanical enhancements in GFRP composites depend on factors like filler-matrix bonding, load transfer, and crack propagation under applied loads. Micro-fillers reinforce low-density structures, while nano-fillers offer diverse reinforcement capabilities. The shape, size, dispersion, and adhesion of both natural and synthetic fillers significantly affect performance. Ongoing research enables tailored GFRP composites for specific needs, advancing multifunctional polymer materials for everyday applications. Particle fillers are increasingly used in GFRPs to enhance mechanical, physical, and chemical properties. As shown in Table 6.1, hybrid composites outperform non-hybridized counterparts. With sustainability in focus, researchers are embracing the "Reduce, Reuse, Recycle" approach—incorporating fillers like SiO_2, WS_2, Si_3N_4, Al_2O_3, LD slag, CNTs, MWCNTs, graphene, and industrial byproducts such as fly ash and eggshell. These additives improve microhardness, wear resistance, and toughness. Marine waste (e.g., clamshells rich in $CaCO_3$) and fly ash cenospheres have shown promise due to their low density, chemical stability, and mechanical reinforcement (Dauphin et al., 2013; Hiremath et al., 2023; Suresha et al., 2008; Deepthi et al., 2010). Clamshells also serve as eco-friendly substitutes for $CaCO_3$ and talc. Figure 6.6 illustrates how wollastonite loading (0–20 wt.%) progressively enhances the mechanical properties of GFRP samples.

Deepthi et al. (2010) and Rajhi et al. (2022) reported improved GFRP performance with additions of nano-silica (NS), MWCNTs, and nano-iron oxide (NFe). SEM analysis of fractured specimens under tensile and flexural loading revealed filler-specific failure modes: (a) no fiber pull-out or breakage, (b) MWCNT agglomeration, and (c) delamination with fiber splitting—highlighting the varied influence of different nano-fillers.

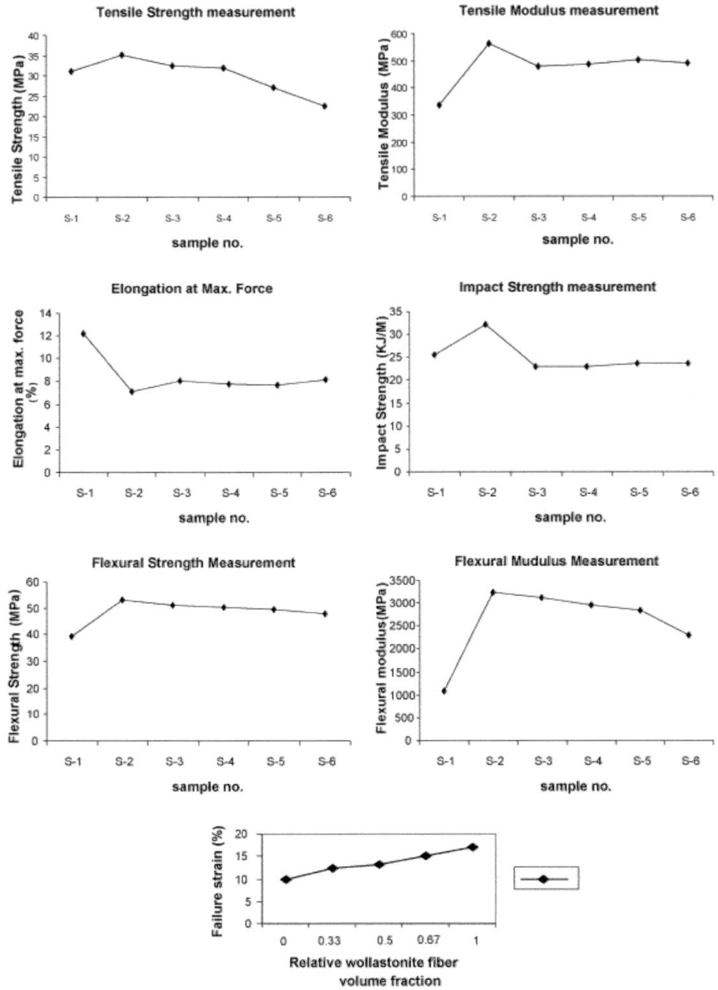

Figure 6.6: Effect of filler in mechanical property of glass fibre composite Himani and Purnima (2010).

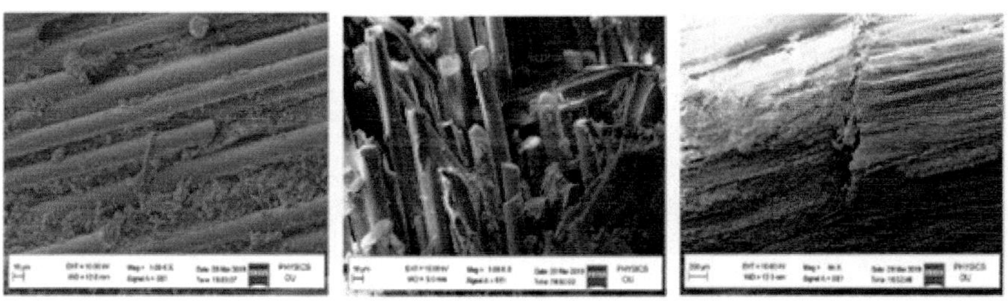

Figure 6.7: SEM image of fractured surface of specimen subjected to tensile load for 0.5 wt.% (a) NS, (b) MWCNTs, and (c) NFe (Rajhi, 2022) (Open Access).

Table 6.1: Mechanical property of hybrid GFRP.

Glass fibre content (Wt.%)	Filler detail	Tensile strength (MPa)	Tensile modulus (GPa)	Flexural strength (MPa)	Flexural modulus (GPa)	ILSS (MPa)	Impact strength (kJ/m^{2})	Micro hardness	References
55	0%	305.4	12.62	334.22	21.23 Mpa	–	1350 J/m	–	(Jena and Katiyar, 2021)
	SiC 5%, 10%	342.1, 355.8	12.78, 12.91	464.13, 475.22	26.58, 26.94 MPa	–	885, 918 J/m	–	
	Graphite 5%, 10%	402.2, 404.2	13.05, 13.11	469.72, 455.84	25.35, 25.33 MPa	–	650, 710 J/m	–	
–	Graphene 0%, 3%, 6%, 9%	141.32, 178.38, 181.55, 178.34	1.321, 1.483, 1.327, 1.183	160.17, 184.09, 188.69, 153.88	11.09, 5.938, 4.965, 4.525		87, 98.04, 110.82, 129.61		(Hulugappa et al., 2016)
45	Chicken eggshell 0%, 3%, 6%	172, 150, 131	4.96, 6.15, 6.63				53.055, 59.874, 75.384		(Gara et al., 2021)
Glass + Kevlar (50%, 49.70%, 49.40%, 49.10%)	MWCNT 0%, 0.3%, 0.6%, 0.9%	135, 150, 178, 170	3407, 3507, 3937, 3210	104.2, 120.2, 154.9, 110.5	3005, 2015, 8055, 5085			18, 20.5, 22.3, 21.2 Hv	(Nayak et al., 2015)
	Linz-Donawitz 0%, 7.5%, 15%, 22.5%	245.54, 221.43, 192.57, 175.21						24.98, 26.31, 29.26, 32.83 Hv	(Rangaswamy et al., 2021)

Table 6.1 contd. ...

...Table 6.1 contd.

Glass fibre content (Wt.%)	Filler detail	Tensile strength (MPa)	Tensile modulus (GPa)	Flexural strength (MPa)	Flexural modulus (GPa)	ILSS (MPa)	Impact strength (kJ/m$^{2)}$	Micro hardness	References
50%	Graphite 0%, 5%	164.51, 191	8558, 10006 MPa					102, 112 H	(Pati and Satapathy, 2015)
50%	Iron mud 0%, 5%, 10%, 15%, 20%	268.57, 141.71, 158.55, 225.03, 244.93	6.67, 8.35, 9.17, 9.68, 9.72	312.435, 207.4887, 216.5852, 289.0973, 300.3977		22.42, 15.49, 15.45, 19.72, 19.98	687.40, 1125.58, 1218.30, 671.01, 552.24 J/cm^2	50, 55, 66, 62, 62 Barcol	(Suresha et al., 2008)
50	SiO_2 0%, 5%, 7.5%, 10%	254, 300, 316.6, 326.7	8.34, 9.43, 9.53, 9.57						(Pani et al., 2019)
	Graphite 0%, 2.5%, 5%, 7.5%	16.51, 182.6, 191, 205	8558, 9857, 10006, 10501 MPa						(Raju et al., 2012)

6.5 Three Body Abrasive Wear of GFRP Composites

Polymer composites reinforced with glass fibers and fillers offer excellent strength-to-weight and stiffness-to-weight ratios, making them ideal for sliding and rolling components such as bearings, gears, seals, and clutches (Almajid et al., 2011). Incorporating various fibers enhances wear resistance in FRPs, as supported by several studies (El-Tayeb, 2008; Bijwe and Rattan, 2007; Soney et al., 2023). Glass fiber, a widely used synthetic fiber, is valued for its thermal stability, availability, and cost-effectiveness. It notably improves wear resistance in polymer composites under sliding and reciprocating motions (Gupta, 2012). However, its inherent brittleness can lead to surface scuffing. To overcome this, particulate fillers are introduced as secondary reinforcements—enhancing fiber–matrix bonding, improving mechanical and tribological properties, and lowering costs—thereby enabling the development of hybrid GFRPs with superior performance (Jena, 2019). Wear refers to surface degradation caused by the relative motion between contacting bodies. Common wear types include corrosive, fretting, abrasive, adhesive, erosive, and surface fatigue. Of these, abrasive wear is the most widespread—responsible for nearly half of all wear issues—occurring when hard particles slide against a surface.

There are two main abrasive modes:

- **Two-body abrasion:** wear is caused by direct contact with hard asperities on the opposing surface.
- **Three-body abrasion:** loose particles between surfaces lead to material loss from both sides.

Figure 6.8 illustrates these mechanisms. Figure 6.9 highlights how wear behavior varies with parameters like load, motion, surface geometry, sliding speed, temperature, humidity, lubrication, and material composition.

In glass-polymer composites, fillers enhance abrasion resistance by strengthening the material and improving its ability to withstand wear and friction (Katiyar, n.d). They also serve as sacrificial elements, absorbing abrasive energy and protecting the matrix and fibers from damage (Selvaraj et al., 2021). High-performance fillers, combined with a protective layer, enhance abrasion resistance by dispersing mechanical loads and minimizing localized wear. This stress-sharing improves durability and reduces wear rates. As shown in Table 6.2, filler selection significantly affects wear behavior, maintaining low friction and controlling abrasive mechanisms—making it crucial to tailor filler choice to specific applications and performance needs (Carrión-Vilches et al., 2015). Filler content significantly influences abrasion resistance in GFRP composites, as shown in Table 6.2. However, excessive filler can raise

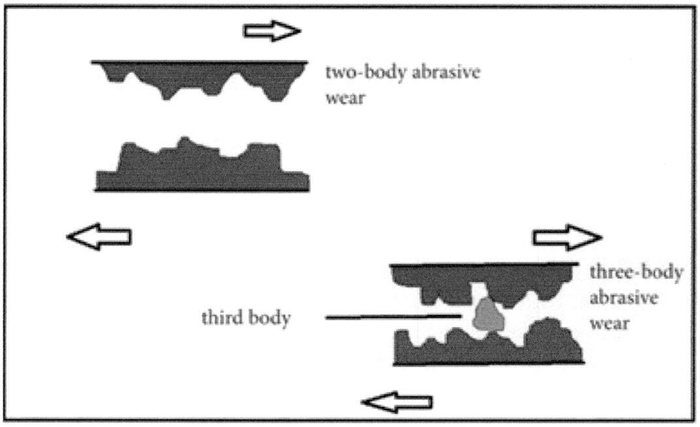

Figure 6.8: Two body and three body abrasive wear (Suresha et al., 2007) (Open Access).

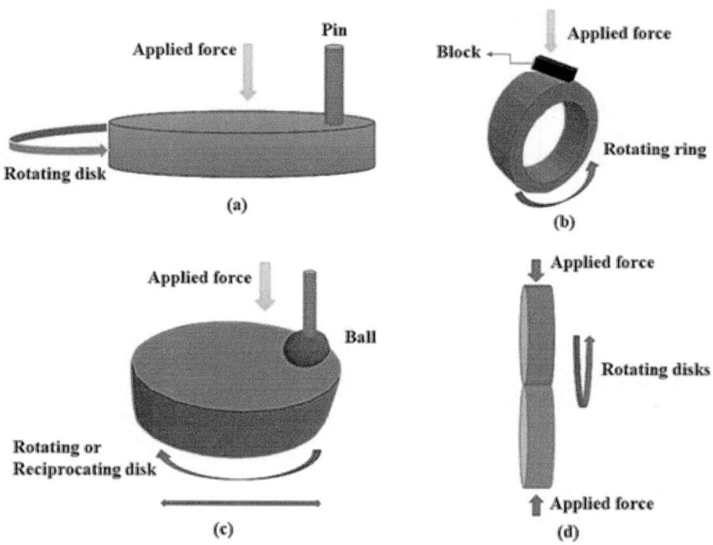

Figure 6.9: Different tribo test (a) Pin on disk (POD), (b) Block on ring, (c) Ball on Disk and (d) Twin disk (TD) tribometer (Mohan et al., 2012).

resin viscosity, complicating processing and affecting mechanical performance. Table 6.2 highlights that combining multiple filler types can yield synergistic improvements in wear resistance and toughness. For instance, Subbaya et al. (2012) demonstrated this by using both synthetic and natural fillers. Surface treatment of fillers is also key to ensuring good matrix bonding and optimizing abrasion resistance.

Table 6.2: Three body abrasion wear test of glass fibre polymer composite.

Matrix	Glass fibre content	Filler content	Tribo-Test	Process parameter	Wear resistance	References
Epoxy L12	Glass + Kevlar (50%, 49.70%, 49.40%, 49.10%)	MWCNT 0%, 0.3%, 0.6%, 0.9%	Sliding wear test	Velocity: 3.5m/s, Load: 40 N, Sliding duration: 1 min	Increase	(Karle and Tungikar, 2021)
Epoxy L12	50%	Graphite 0%, 5%	Rubber wheel abrasion test	Rotational speed: 200 RPM, Abrasive feed rate: 255 g/min	Highest wear obtained for grahite filled composite	(Kuram, 2019)
Epoxy L12	50 wt%	Iron mud 0%, 5%, 10%, 15%, 20%	Dry rubber wheel abrasion test	Erodent flow rate: 363 g/min	Addition of iron mud improved wear resistance under controlled parameters	(Seeli et al., 2023)
Epoxy L12	50 wt%	SiO_2 0%, 5%, 7.5%, 10%	Dry rubber wheel abrasion test	Abrading distance: 250–1000 m, Load: 24 and 36 N	Reduction in steady wear rate observed for composites with 10 wt.% SiO_2	(Ferreira et al., 2010)
Polyester resin (Neoxil CE92)	Glass + aramid		Block on shaft wear test	Speed: 500,710 rpm Load: 500, 1000 g	Aramid fiber mixed composites undergo less wear	(Bhat et al., 2023)
Epoxy L12		Graphite 0%, 2.5%, 5%, 7.5%	Pin on disk	Load: 40 N, Velocity: 4 m/s, Sliding distance: 500–6000 m		(Pıhtılı and Tosun, 2002)
Epoxy L12	Glass + Carbon		Three-body abrasive wear	Abrasive feed rate: 363 g/min, Speed of rubber wheel: 1.5 m/s	GCG_2C shows least specific wear rate of 18.8476×10^{-3} mm^3Nm^{-1}	(Suresha et al., 2007)
Unsaturated Polyester	Short Fibre/30 mm/–40,45 wt.%	CaCO (5 wt.%)	Wheel Abrasion Test	Load: 3, 4.5, 6, 7.5 N, Sliding speed: 0.83, 1.66, 2.5, 3.34, 4.16 m/s, Sliding distance: 100, 200, 300, 400, 500, 600, 700, 800 m		(Jesthi and Nayak, 2020)
Epoxy	Bidirectional plain woven fibre, 40, 50, 60 vol.%	Graphite (10 wt.%), SiC (10 wt.%)	Pin-on-disc test	Load: 20, 40, 60 N, Sliding velocity: 3,4,5 m/s, Sliding distance: 1000, 2000, 3000 m	Increased	(Chand et al., 2000)

Solid fillers like graphite and graphite oxide enhance GFRP hardness while acting as lubricants to reduce wear—though excessive use can cause flaking (Basavarajappa et al., 2009). Graphite particles embedded in polymer coatings are also effective in lowering friction and wear under controlled conditions (Gafsi et al., 2021). Cost-effective filler strategies include using metallic/non-metallic powders from industrial and agricultural waste such as red mud, fly ash, copper slag, cenospheres, eggshells, feathers, rice husks, and seashells—all of which have shown promising results in GFRPs. Numerous studies affirm that secondary fillers significantly enhance the dry abrasion resistance of GFRPs (Rout and Satapathy, 2012; Kumar and Velmurugan, 2020; Su et al., 2005; Katiyar and Hena, 2023). Biswas and Satapathy (2009) demonstrated positive effects using red mud, an industrial waste, as a filler. Overall, the use of fillers can greatly improve abrasion resistance in glass-polymer composites. However, the filler type, content, and distribution must be carefully optimized to maintain mechanical integrity.

6.6 Summary

The chapter highlights how fillers influence the mechanical and abrasive wear behavior of glass fiber composites. Fillers improve load distribution, boost load-bearing capacity, and enhance resistance to wear under reciprocating loads. Their impact varies with type, concentration, and operating conditions. Proper selection and optimization are essential for achieving desired tribological performance. Fillers also affect heat tolerance and wear behavior at different sliding speeds. However, this study does not deeply examine filler–matrix interface interactions or the effects of processing on composite properties.

6.7 Future Work

- Filler particle size and distribution influence both mechanical and tribological performance and warrant further study.
- Future filler designs may combine nano and micro fillers, include chemical modifications, or adopt self-healing and bio-inspired features.
- Additional tribological aspects such as fatigue, diffusion, erosion, and corrosion should also be explored.

References

Agarwal, G., Patnaik, A. and Sharma, R.K. 2013. Thermo-mechanical properties of silicon carbide-filled chopped glass fiber-reinforced epoxy composites. *International Journal of Anvanced Structural Engineering*, pp. 1–8.

Agrawal, A., Chandraker, S. and Sharma, A. 2018. Physical, mechanical and sliding wear behavior of solid glass microsphere filled epoxy composites. *Materials Today: Proceedings*, 29(2): 420–426.

Almajid, A., Friedrich, K., Floeck, J. and Burkhart, T. 2011. Surface damage characteristics and specific wear rates of a new continuous carbon fibre (CF)/polyetheretherketone (PEEK) composite under sliding and rolling contact conditions. *Appl. Compos. Mater.*, 18(3): 211–30.

Arjmandi, R., Hassan, A., Majeed, K. and Zakaria, Z. 2015. Rice husk filled polymer composites. *International Journal of Polymer Science*, 2015.

Ascione, Francesco, Gabriella Maselli and Antonio Nesticò. 2024. Sustainable materials selection in industrial construction: A life-cycle based approach to compare the economic and structural performances of glass fibre reinforced polymer (GFRP) and steel. *Journal of Cleaner Production*, 475: 143641.

Asrar Rafiq Bhat, Rajiv Kumar, Prasanna Kumar S. Mural. 2023. Natural fiber reinforced polymer composites: A comprehensive review of Tribo-Mechanical properties. *Tribology International*, 189: 2023. https://doi.org/10.1016/j.triboint.2023.108978.

Ates, B., Koytepe, S., Balcioglu, S., Ulu, A., Gurses, C. et al. 2017. Biomedical applications of hybrid polymer composite materials. *Hybrid Polymer Composite Materials: Applications*, pp. 343–408.

Ayswarya, E.P., Nair, A.B. and Thachil, E.T. 2021. A comparative study of mechanical, dynamic mechanical and thermal properties of rice husk ash, modified rice husk ash and nano silica filled epoxy composites. *Materials Today: Proceedings*, 47(15): 5351–5357.

Basavarajappa, S., Arun, K.V. and Paulo Davim, J. 2009. Effect of filler material on dry slidingwear behaviour of polymer matrix composites—A Taguchi approach. *J. Miner. Mater. Character Eng.*, 8: 379–391.

Bijwe, J. and Rattan, R. 2007. Influence of weave of carbon fabric in polyetherimide composites in various wear situations. *Wear*, 263: 984–991.

Bijwe, J., Rajesh, J.J., Jeyakumar, A., Ghosh, A. and Tewari, U.S. 2000. Influence of solid lubricants and fibre reinforcement on wear behaviour of polyethersulphone. *Tribology International*, 34(10): 697–706.

Biswas, S. and Satapathy, A. 2009. Tribo-performance analysis of red mud filled glass-epoxy composites using Taguchi experimental design. *Materials & Design*, 30(8): 2841–2853.

Carrión-Vilches, F.J., González-Vivas, A., Martínez-Mateo, I.J. and Bermúdez M.D. 2015. Study of the abrasion resistance under scratching of polybutyleneterefalate–glass fibre composites. *Tribology International*, 92: 365–378.

Cetin, M.S. and Karahan Toprakci, H.A. 2021. Flexible electronics from hybrid nanocomposites and their application as piezoresistive strain sensors. *Composites Part B: Engineering*, 224.

Chand, N., Naik, A. and Neogi, S. 2000. Three-body abrasive wear of short glass fibre polyester composite. *Wear*, 242: 38–46.

Czech, K., Oliwa, R., Krajewski, D., Bulanda, K., Oleksy, M. et al. 2021. Hybrid polymer composites used in the arms industry: A review. *Materials*, 14(11): 3047.

Das, S. Chandra and MdEH Nizam. 2014. Applications of fiber reinforced polymer composites (FRP) in civil engineering. *International Journal of Advanced Structures and Geotechnical Engineering*, 3.3: 299–309.

Das, Tushar Kanti, Prosenjit Ghosh and Narayan Ch Das. 2019. Preparation, development, outcomes, and application versatility of carbon fiber-based polymer composites: A review. *Advanced Composites and Hybrid Materials*, 2: 214–233.

Dauphin, Y., Ball, A.D., Castillo-Michel, H., Chevallard, C., Cuif, J.P. et al. 2013. *In situ* distribution and characterization of the organic content of the oyster shell Crassostrea gigas (Mollusca Bivalvia). *Micron.*, 44: 373–383.

Deepthi, M.V., Sharma, M., Sailaja, R.R.N., Anantha, P., Sampathkumaran, P. et al. 2010. Mechanical and thermal characteristics of high-density polyethylene-fly ash Cenospheres composites. *Materials & Design*, 31(4): 2051–2060.

Dehrooyeh, S., Vaseghi, M., Sohrabian, M. and Sameezadeh, M. 2021. Glass fiber/Carbon nanotube/Epoxy hybrid composites: Achieving superior mechanical properties. *Mechanics of Materials*, 161.

Del Vecchio, Ciro, Marco Di Ludovico and Andrea Prota. 2021. Cost and effectiveness of fiber-reinforced polymer solutions for the large-scale mitigation of seismic risk in reinforced concrete buildings. *Polymers*, 13.17: 2962.

Dhawan, Vikas, Singh and Sehijpal, Singh, Inderdeep. 2013. Effect of natural fillers on mechanical properties of GFRP composites. *Journal of Composites*, 792620: 8. https://doi.org/10.1155/2013/79262

Edwin Raja Dhas, J. and Arun, M. 2022. A review on development of hybrid composites for aerospace applications. *Materials Today Proceedings*, 64(1): 267–273.

Elsheikh, M.N. and Yakoub, N.G. 2020. Effect of nano Al2O3 addition on the mechanical properties of GFRP reinforced by Mwcnts. *International Journal of Scientific & Technology Research*, 9(9).

El-Tayeb, N.S.M. 2008. A study on the potential of sugarcane fibres/polyester composite for tribological applications. *Wear*, 265(1-2): 223–235.

Espert, A., Vilaplana, F. and Karlsson, S. 2004. Comparison of water absorption in natural cellulosic fibres from wood and one-year crops in polypropylene composites and its influence on their mechanical properties. *Composites Part A: Applied Science & Manufacturing*, 35(11): 1267–1276.

Ferreira, J.A.M., Capela, C. and Costa, J.D. 2010. A study of the mechanical behaviour on fibre reinforced hollow microspheres hybrid composites. *Composites Part A: Applied Science and Manufacturing*, 41(3): 345–352. https://doi.org/10.1016/j.compositesa.2009.10.018.

Gafsi, N., Smaoui, I., Verdejo, R., Kharrat, M., Manchado, M.Á.L. et al. 2021. Tribological and mechanical characterization of epoxy/graphite composite coatings: Effects of particles' size and oxidation. *Proceedings of the Institution of Mechanical Engineers, Part J: Journal of Engineering Tribology*, 235(1): 129–137.

Ganeshkumar, S. 2024. Multifunctional polymer composites: Design, properties, and emerging applications—A critical review. *International Conference on Eco-friendly Fibers and Polymeric Materials*. Singapore: Springer Nature Singapore.

Gara, D.K., Raghavendra, G. and Prasad, P.S., S O. 2021. Enhanced mechanical properties of glass fibre epoxy composites by 2D exfoliated graphene oxide filler. *Ceramics International*, 47(24): 34860–34868.

Genovés, V., Fariñas, M.D., Pérez-Aparicio, R., Saiz-Rodríguez, L., Valentín, J.L. et al. 2022. Micronized recycle rubber particles modified multifunctional polymer composites: Application to ultrasonic materials engineering. *Polymers*, 14(17): 3617.

Gu, H., Liu, C. and Zhu, J. 2018. Introducing advanced composites and hybrid materials. *Adv. Compos. Hybrid Mater.*, 1: 1–5. https://doi.org/10.1007/s42114-017-0017-y.

Harussani, M.M., Sapuan, S.M., Nadeem, G., Rafin, T., Kirubaanand, W. et al. 2022. Recent applications of carbon-based composites in defence industry: A review. *Defence Technology*, 18(8): 1281–1300.

Hepsiba Seeli, V.V.S. Prasad and Maduthuri Venkatesh. 2023. Nano-graphene E-glass fibre epoxy polymer composite thermal and mechanical characteristics study. *Materials Today: Proceedings*. https://doi.org/10.1016/j.matpr.2023.03.341.

Hiremath, A., Nayak, S.Y., Heckadka, S.S. and Pramod, J.J. 2024. Mechanical behavior of basalt-reinforced epoxy composites modified with biomass-derived seashell powder. *Biomass conversion and Biorefinery*, 14: 26281–26291.

Hulugappa, B., Achutha, M.V. and Suresha, B. 2016. Effect of fillers on mechanical properties and fracture toughness of glass fabric reinforced epoxy composites. *Journal of Minerals and Materials Characterization and Engineering*, 04: 1–14.

Jena, H. 2019. Study of tribo-performance and application of polymer composite. *In*: Katiyar, J., Bhattacharya, S., Patel, V. and Kumar, V. (eds.). Automotive Tribology. Energy, Environment, and Sustainability. Springer, Singapore. https://doi.org/10.1007/978-981-15-0434-1_5.

Jena, H. and Katiyar, J.K. 2021. Tribo-analysis of polymer composite in spur gear. *In*: Hameed Sultan, M.T., Mohd Jamir, M.R., Abdul Majid, M.S., Azmi, A.I. and Saba, N. (eds.). *Tribological Applications of Composite Materials. Composites Science and Technology*. Springer, Singapore. https://doi.org/10.1007/978-981-15-9635-3_12.

Jesthi, D.K. and Nayak, R.K. 2020. Influence of glass/carbon fiber stacking sequence on mechanical and three-body abrasive wear resistance of hybrid composites. *Materials Research Express*, 7.

Jitendra Kumar Katiyar and Hemalata Jena. 2023. 6—The role of synthetic fibers in the tribological behavior of polymeric materials. pp. 173–194. *In*: Soney C. George, Jozef T. Haponiuk, Sabu Thomas, Rakesh Reghunath and Sarath, P.S. (eds.). *Elsevier Series on Tribology and Surface Engineering, Tribology of Polymers, Polymer Composites, and Polymer Nanocomposites*, Elsevier. https://doi.org/10.1016/B978-0-323-90748-4.00002-9.

Joshi Himani, J. Purnima. 2010. Development of glass fiber, wollastonite reinforced polypropylene hybrid composite: Mechanical properties and morphology. *Materials Science and Engineering: A*, 527(7-8): 1946–1951, ISSN 0921-5093. https://doi.org/10.1016/j.msea.2009.11.039.

Karle, A.H. and Tungikar, V.B. 2021. Optimization of wollastonite reinforced epoxy composites. *Materials Today: Proceedings*, 45(6): 5153–5157.

Karnati, S.R., Agbo, P. and Zhang, L. 2020. Applications of silica nanoparticles in glass/carbon fiber-reinforced epoxy nanocomposite. *Composites Communications*, 17: 32–41.

Kenyó, C., Renner, K., Móczó, J., Fekete, E., Kröhnke, C. et al. 2018. Hips/zeolite hybrid composites as active packaging materials: Structure and functional properties. *European Polymer Journal*, 103: 88–94.

Knippers, J. and Waimer, F. 2015. A novel hybrid composite construction for complex concrete shells in architecture. *(IASS) Symposium, Amsterdam, The Netherlands*, 11: 1–11.

Kowshik, S., Shettar, M., Rangaswamy, N., Chate, G., Somdee, P. et al. 2022. Effect of nanoclay on mechanical, flammability, and water absorption properties of glass fiber-epoxy composite. *Cogent Engineering*, 9(1).

Kumar, M. and Velmurugan, V. 2020. Enhancement of mechanical and tribological properties of epoxy based polymer with TALC & PTFE as fillers. *Transstellar Journal Publications and Research Consultancy Private Limited*, 10(3): 137–146.

Kumar, Manoj, Hemalata Jena and Priyabrat Pradhan. 2025. Impact of filler on solid particle erosion wear in glass fibre composites: A comprehensive review. *Proceedings of the Institution of Mechanical Engineers, Part E: Journal of Process Mechanical Engineering*, 09544089251326998.

Kuram, E. 2019. Hybridization effect of talc/glass fiber as a filler in polycarbonate/acrylonitrile-butadiene-styrene composites. *Composites Part B: Engineering*, 173.

Lazar, Prince Jeya Lal. 2024. Application of metaheuristics to evaluate energy absorption and damage tolerance in GFRP-epoxy nanocomposite laminates. *International Journal on Interactive Design and Manufacturing (IJIDeM)*, 1–17.

León Gómez, J.C., De León Aldaco, S.E. and Aguayo Alquicira, J. 2023. A review of hybrid renewable energy systems: architectures, battery systems, and optimization techniques. *Eng.*, 4(2): 1446–1467.

Madhavan, M.K., Sathyan, D. and Jayanarayanan, K. 2021. Hybrid natural fiber composites in civil engineering applications. *Hybrid Natural Fiber Composites: Material Formulations, Processing, Characterization, Properties, and Engineering Applications*, pp. 41–72.

Majeed, K., Jawaid, M., Hassan, A., Abu Bakar, A., Abdul Khalil, H.P.S. et al. 2013. Potential materials for food packaging from nanoclay/natural fibres filled hybrid composites. *Materials & Design*, 46(1): 391–410.

Mohammadi, Hossein. 2022. Lightweight glass fiber-reinforced polymer composite for automotive bumper applications: A review. *Polymers*, 15.1: 193.

Mohan, N., Natarajan, S. and Kumareshbabu, S.P. 2012. The role of synthetic and natural fillers on three-body abrasive wear behaviour of glass fabric-epoxy hybrid composites. *Journal of Applied Polymer Science*, 124(1): 484–494.

Mohanty, Saswatam. 2024. Enhancing mechanical properties of basalt fibre composites with clamshell waste from sea bed. *The Journal of Solid Waste Technology and Management*, 50.3: 532–542.

Morin, Brian. 2009. Weight reduction and cost savings using hybrid composites containing high modulus polypropylene fiber. *Composites and Polycon*, 15–17.

Nagaraja, K.C., Ranjana, S. and Prakash, S.G. 2018. Effect of rubber powder as filler on Glass fiber reinforced epoxy composites. *International Journal of Management, Technology and Engineering*, 8(11): 21–27.

Nayak, S.Y., Heckadka, S.S., Shrma, P., Aman, I., Dey, S. et al. 2015. Use of egg shell particulate as fillers in E-glass/epoxy composites. *Proceedings of International Conference on Mechanical Engineering and Industrial Automation*, pp. 21–25.

Neşer, G. 2017. Polymer based composites in marine use: History and future trends. *Procedia Engineering*, 194: 19–24.

Nwobi-Okoye, Chidozie Chukwuemeka, Martins Kenechukwu Anyichie and Clement Uche Atuanya. 2020. RSM and ANN modeling for production of newbouldia laevies fibre and recycled high density polyethylene composite: Multi objective optimization using genetic algorithm. *Fibers and Polymers*, 21: 898–909.

Palanikumar, K., Latha, B., Senthilkumar, V.S. and Davim, J.P. 2012. Analysis on drilling of glass fiber-reinforced polymer (GFRP) composites using grey relational analysis. *Materials and Manufacturing Processes*, 27(3): 297–305.

Pani, B., Chandrasekhar, P. and Singh, S. 2019. Application of box-behnken design and neural computation for tribo-mechanical performance analysis of iron-mud-filled glass-fiber/epoxy composite and parametric optimization using PSO. *Polymer Composites*, 40(4): 1433–1449.

Pati, P.R. and Satapathy, A. 2015. Processing and characterization of glass-epoxy composites filled with Linz-donawitz (LD) slag. *Universal Journal of Mechanical Engineering*, 3(1): 7–11.

Petrény, R., Tóth, C., Horváth, A. and Mészáros, L. 2022. Development of electrically conductive hybrid composites with a poly(lactic acid) matrix, with enhanced toughness for injection molding, and material extrusion-based additive manufacturing. *Heliyon*, 8(8).

Pıhtılı, H.H. and Tosun, N. 2002. Effect of load and speed on the wear behaviour of woven glass fabrics and aramid fibre-reinforced composites. *Wear*, 252(11-12): 979–984.

Puttegowda, M., Rangappa, S.M., Jawaid, M., Shivanna, P., Basavegowda, Y. et al. 2018. Potential of natural/synthetic hybrid composites for aerospace applications. *Sustainable Composites for Aerospace Applications*, pp. 315–351.

Raja, V.L., Muralidharan, K. and Dhanasekaran, R. 2020. Prediction of wear characteristics of polymer composites by ANN modified by GA. *Recent Trends in Mechanical Engineering: Select Proceedings of ICIME 2020*. Singapore: Springer Singapore, 231–237.

Rajhi, A.A. 2022. Mechanical characterization of hybrid nano-filled glass/epoxy composites. *Polymers*, 14: 4852. https://doi.org/10.3390/polym14224852.

Raju, B.R., Suresha, B., Swamy, R.P. and Bharath, K.N. 2012. The effect of silicon dioxide filler on the wear resistance of glass fabric reinforced epoxy composites. *Polymer Science and Technology*, 2(4): 51–57.

Ramkumar Yadav, Hwalim Lee, Jung-Hwan Lee, Rajendra K. Singh, Hae-Hyoung Lee, A. et al. 2023. Comprehensive review: Physical, mechanical, and tribological characterization of dental resin composite materials. *Tribology International*, 179. https://doi.org/10.1016/j.triboint.2022.108102.

Rangaswamy, H.M.H.H., Chandrashekarappa, M.P.G., Pimenov, D.Y., Giasin, K., Wojciechowski, S. et al. 2021. Experimental investigation and optimization of compression moulding parameters for MWCNT/glass/kevlar/epoxy composites on mechanical and tribological properties. *Journal of Materials Research and Technology*, 15: 327–341.

Ravishankar, B., Nayak, S.K. and Kader, M.A. 2019. Hybrid composites for automotive applications—A review. *Journal of Reinforced Plastics and Composites*, 38(18): 835–845.

Reddy, S.S.P., Suresh, R., Hanamantraygouda, M.B. and Shivakumar, B.P. 2021. Use of composite materials and hybrid composites in wind turbine blades. *Materials Today: Proceedings*, 46(7): 2827–2830.

Rout, A. and Satapathy, A. 2012. Analysis of dry sliding wear behaviour of rice husk filled epoxy composites using design of experiment and ANN. *Procedia Engineering*, 38: 1218–1232.

Selvaraj, Senthil Kumaran, Raj, Aditya, Dharnidharka, Mohit, Chadha, Utkarsh et al. 2021. Cutting-edge survey of tribological behavior evaluation using artificial and computational intelligence models. *Advances in Materials Science and Engineering*, 9529199: 17. https://doi.org/10.1155/2021/9529199.

Seydibeyoğlu, M.Ö., Dogru, A., Wang, J., Rencheck, M., Han, Y. et al. 2023. Review on hybrid reinforced polymer matrix composites with nanocellulose, nanomaterials, and other fibers. *Polymers*, 15(4): 984.

Shivanagere, Abhay, Sharma, S. and Goyal, P. 2018. Modelling of glass fibre reinforced polymer (Gfrp) for aerospace applications. *Journal of Engineering Science and Technology*, 13.11: 3710–3728.

Siddhartha, Gupta K. 2012. Mechanical and abrasive wear characterization of bidirectional and chopped E-glass fibre reinforced composite materials. *Mater. Des*., 35: 467–479.

Singh, B., Gupta, M. and Anchal Verma. 2003. Polyester moulding compounds of natural fibres and wollastonite. *Composites Part A: Applied Science and Manufacturing*, 34(11): 1035–1043, https://doi.org/10.1016/S1359-835X(03)00239-2.

Singh, K.K. and Kumar, S. 2021. Tribological performance of graphene nanoplatelets filled glass/epoxy composites under dry, inert gas and oil-lubricated environmental conditions. *Materials Letters*, 282.

Su, F.H., Zhang, Z.Z. and Liu, W.M. 2005. Study on the friction and wear properties of glass fabric composites filled with nano- and micro-particles under different conditions. *Materials Science and Engineering: A*, 392(1-2): 359–365.

Subbaya, K., Rajendra, N., Varadarajan, Y. and Suresha, B. 2012. Multiple response optimization of three-body abrasive wear behaviour of graphite filled carbon-epoxy composites using grey-based Taguchi approach. *Journal of Minerals and Materials Characterization and Engineering*, 11: 876–884. doi: 10.4236/jmmce.2012.119082.

Suhot, M.A., Hassan, M.Z., Aziz, S.A., Md Daud, M.Y. 2021. Recent progress of rice husk reinforced polymer composites: A review. *Polymers*, 13(15): 2391.

Suresha, B., Chandramohan, G., Renukappa, N.M. and Siddaramaiah, , Jayaraju T. 2007. Mechanical and tribological properties of glass-epoxy composites with and without graphite particulate filler. *Journal of Applied Polymer Science*, 103(4): 2472–2480.

Suresha, B., Chandramohan, G. and Siddaramaiah, Jayaraju T. 2008. Three-body abrasive wear behavior of E-glass fabric reinforced/graphite- filled epoxy composites. *Polymer Composites*, 29(6): 631–637.

Suresha, B., Chandramohan, G. and Siddaramaiah, Jayaraju, T. 2008. Influence of cenosphere filler additions on the three-body abrasive wear behavior of glass fiber-reinforced epoxy composites. *Polymer Composites*, 29(3): 307–312.

Thakur, V.K., Thakur, M.K. and Pappu, A. 2017. *Hybrid Polymer Composite Materials Properties and Characterisation*. Woodhead Publishing.

Trzepieciński, T., Batu, T., Kibrete, F. and Lemu, H.G. 2023. Application of Composite materials for energy generation devices. *Journal of Composites Science*, 7(2): 55.

Turaka, S. and Bandaru, A.K. 2023. Enhancement in mechanical properties of glass/epoxy composites by a hybrid combination of multi-walled carbon nanotubes and graphene nanoparticles. *Polymers*, 15(5).

Wan Badaruzzaman, W.H., Dabbagh, N.M.R., Salleh, K.M., Saharuddin, E.N., Mat Radzi, N.F. et al. 2022. Mechanical properties and water absorption capacity of hybrid GFRP composites. *Polymers*, 14: 1394. https://doi.org/10.3390/polym14071394.

Wang, L. 2020. Fatigue behaviour of composite sandwich beams strengthened with GFRP stiffeners. *Engineering Structures*, 214: 110596.

Wibawa, Tri. 2024. Enhancing the mechanical and fire-resistant properties of GFRP composite using boric acid and sodium silicate fillers. *Engineering, Technology & Applied Science Research*, 14.6: 18911–18922.

Yung, K.C., Zhu, B.L., Yue, T.M. and Xie, C.S. 2009. Preparation and properties of hollow glass microsphere-filled epoxy-matrix composites. *Composites Science and Technology*, 69(2): 260–264.

Zamora-Mendoza, L., Gushque, F., Yanez, S., Jara, N., Álvarez-Barreto, J.F. et al. 2023. Plant fibers as composite reinforcements for biomedical applications. *Bioengineering*, 10(7).

Zhang, L. 2015. *The Application of Composite Fiber Materials in Sports Equipment*. Atlantis Press.

CHAPTER

7 Advances in Laser Cladding

Material Trends, Applications, and Future Perspectives

Vikas Diwakar,[1,*] *Sunil kumar Yadav*,[2] *Mohd Zaheer Khan Yusufzai*[1] and *Meghanshu Vashista*[1]

Laser cladding (LC) is a surface modification technique that uses a laser as a heat source to precisely melt depositing materials, resulting in minimal thermal distortion. This capability makes LC suitable for a wide range of fields, including automotive, aerospace, healthcare and energy. Selecting the appropriate material for LC is critical, as it directly influences the performance of the coated surfaces. HEAs—comprising five or more principal elements in near-equiatomic ratios—offer outstanding strength, thermal stability, and corrosion resistance. Similarly, functionally graded materials (FGMs) enable engineers to tailor property gradients for specific service conditions. In one case study, FeCoCrNiMo HEA powder was deposited onto a steel substrate. By optimizing the powder feed rate (*PFR*), scanning speed (*Ss*), and laser power (*Lp*), a uniform, defect-free multi-track coating was achieved.

[1] Department of Mechanical Engineering, Indian Institute of Technology (BHU) Varanasi, Varanasi, India.

[2] Madan Mohan Malaviya University of Technology, Gorakhpur, India.

* Corresponding author: vikasdiwakar.rs.mec21@itbhu.ac.in

ORCID IDs of authors: https://orcid.org/0000-0001-7257-8168; https://orcid.org/0009-0009-4984-8129; https://orcid.org/0000-0001-6514-2497; https://orcid.org/0000-0002-4168-7406

DOI: 10.1201/9781003607359-7

Microstructural analysis revealed fine dendritic growth and a significant increase in microhardness compared to the base metal. Despite its benefits, LC presents several challenges, including high initial equipment costs, material compatibility issues, and the need for precise process optimization. Overcoming these barriers is crucial for further technological progress. Future developments are expected to focus on hybrid coating systems, real-time diagnostics, and the use of sustainable powders and energy-efficient lasers to enhance both the economic and environmental viability of LC.

7.1 Introduction

Laser cladding (LC) has emerged as a leading surface engineering and additive manufacturing technique to enhance material performance, durability, and functionality across diverse industries. Its precision enables high-quality coating deposition and complex geometry fabrication with minimal thermal damage (Imran et al., 2024). Once a niche method, LC is now widely adopted due to advances in laser systems, materials, and automation (Xiang et al., 2024). This chapter outlines emerging material trends, industry-specific applications, and future developments driving LC's expanding role in advanced manufacturing.

7.1.1 Overview of LC Technology

Laser cladding (LC) uses a high-energy laser to melt and bond powder or wire feedstock onto a substrate, forming a dense, low-porosity coating with strong adhesion and minimal heat-affected zone (Rong et al., 2025). Its precision and flexibility surpass traditional methods like thermal spraying or welding, enabling both surface enhancement and additive manufacturing. Originating in the 1970s (Jin et al., 2024), LC has evolved through advancements in fiber/diode lasers, materials, and automation, becoming a key technology for improving component performance, reducing waste, and extending service life.

7.1.2 Importance of LC in Modern Manufacturing

Recent emphasis on sustainable manufacturing has driven growing demand for high-performance materials. Laser cladding (LC) offers a versatile and eco-friendly solution, enabling deposition of tailored coatings with enhanced corrosion, wear, or thermal resistance (Yuan et al., 2024). This makes LC especially effective for components operating under extreme service conditions across various industries.

Laser cladding (LC) enables efficient repair of high-value components by restoring worn parts to their original dimensions, reducing both material waste and replacement costs (Niu et al., 2024). This restoration process enhances service life and promotes resource conservation (Zhang et al., 2023).

7.1.3 Scope and Objectives

The chapter outlines key developments in laser cladding (LC), focusing on three main areas: material trends, applications, and future outlook. It begins by covering LC fundamentals and processing techniques, followed by an exploration of materials such as metallic alloys, ceramics, composites, and functionally graded materials (FGMs), each selected for their unique properties. The applications section highlights LC's integration across sectors including energy, biomedical, aerospace, and defense. In the case study, HEA was successfully cladded onto the substrate, resulting in enhanced hardness and refined microstructure. The chapter concludes by outlining future directions in LC, including hybrid techniques, AI integration, novel material development, and emerging applications such as space exploration.

7.2 Fundamentals of LC

Laser cladding (LC) is a surface engineering technique that uses a high-powered laser to deposit a material layer onto a substrate, enhancing properties such as wear and corrosion resistance. Its effectiveness relies on combining key parameters to ensure precise, durable coatings.

7.2.1 LC Process

Laser cladding (LC) involves melting feedstock material at the interaction zone between the laser beam and the substrate surface, forming a metallurgically bonded layer. As illustrated in Fig. 7.1, the laser—typically fiber, diode, or CO_2—serves as the primary heat source, precisely delivering energy to melt both the substrate surface and the feedstock (Zheng et al., 2025; Xie et al., 2025; Diwakar et al., 2024). Fiber lasers, in particular, offer higher efficiency and superior coating quality. Optical systems focus the beam for accurate energy distribution, ensuring uniform deposition and minimal dilution. In laser cladding, feedstock —either powder or wire—is delivered to the laser–substrate interaction zone. In powder-based LC, a carrier gas (e.g., argon or nitrogen) transports the particles through a nozzle, where nozzle design and gas flow ensure uniform deposition. In wire-based LC, the wire is directly melted into the molten pool by the laser, making it ideal for high-deposition-rate applications such as large part repairs or thick coatings.

During laser cladding, a molten pool forms where the feedstock, substrate, and laser interact—creating a strong metallurgical bond. Rapid solidification yields a dense, low-porosity coating with minimal base material dilution (Yamaguchi et al., 2023; Yamaguchi et al., 2025). Bond quality depends on laser power, scan speed, and material compatibility. To build multilayer or 3D structures, the laser and feed system move relative to the substrate. Key

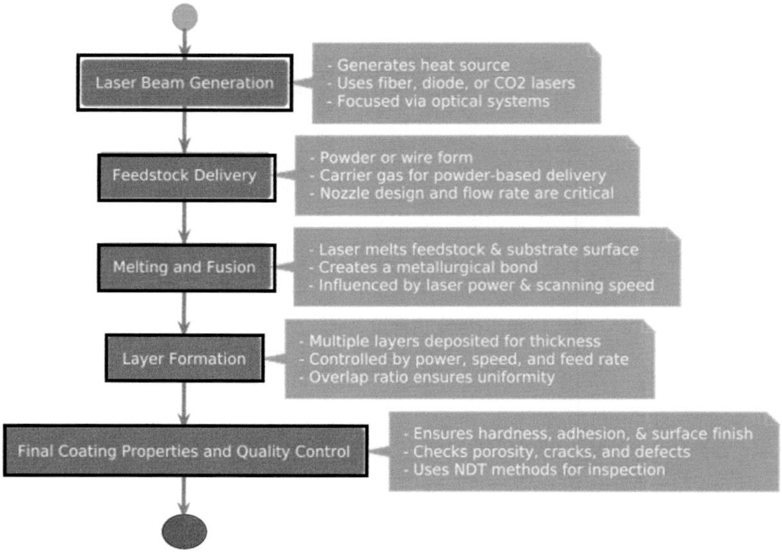

Figure 7.1: LC process mechanism involved for deposition on substrate.

parameters—powder feed rate (PFR), scanning speed (Ss), and laser power (Lp)—must be precisely controlled to achieve the desired thickness, microstructure, and properties (Hao et al., 2023; Li et al., 2022). An optimal overlap between tracks ensures coating uniformity and minimizes surface defects.

The LC process relies on several components, each playing an important role in ensuring the quality and efficiency of the deposition. A systematic flow diagram is shown in Fig. 7.2, which lists all components involved in LC.

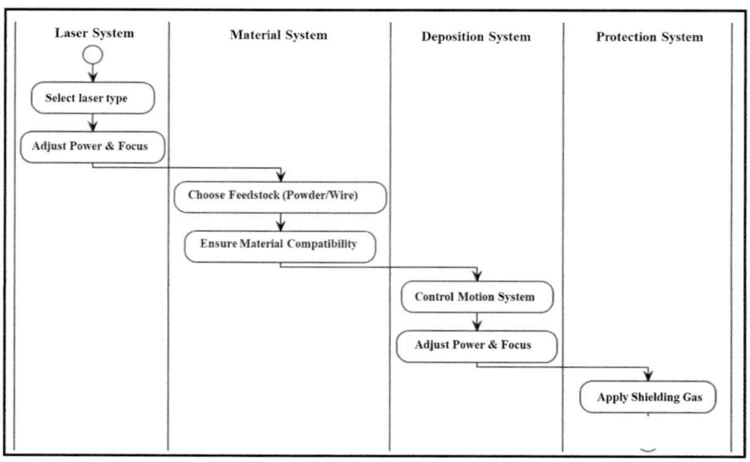

Figure 7.2: Different types of LC systems.

7.2.2 Types of LC Techniques

Laser cladding (LC) can be performed using various techniques tailored to specific materials and applications. As shown in Fig. 7.3, the most common methods are categorized by feedstock type—primarily powder-based and wire-based—each offering distinct advantages in deposition precision and rate. Powder-based laser cladding is the most widely used method due to its precise control over layer composition and microstructure. By tuning laser power, scan speed, and powder feed rate, engineers can achieve high-quality coatings—making it ideal for precision applications.

In wire-based laser cladding (LC), metallic wire is directly fed into the laser-generated melt pool, enabling high-deposition rates and effective thick coatings. It offers advantages over powder-based methods, such as lower cost and reduced material waste, though it requires precise control of wire feed and positioning for consistent results (Zhao et al., 2021). Hybrid LC integrates other deposition processes—like arc welding or plasma spraying—to boost deposition rates, improve material properties, and reduce costs. For instance, laser-arc hybrid cladding provides enhanced process stability and higher productivity, expanding its application for manufacturing and refurbishing complex components.

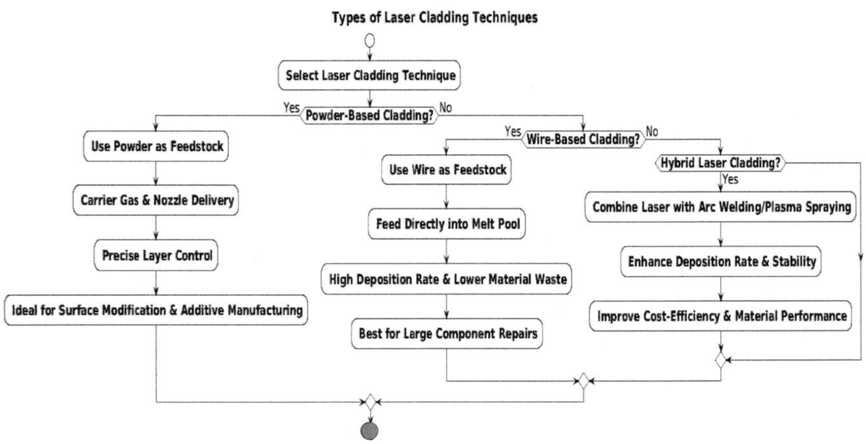

Figure 7.3: LC technique based on feedstock materials.

7.3 Material Trends in LC

The success of laser cladding (LC) largely hinges on the materials selected. Advances in tailored materials have enabled coatings with superior wear, corrosion, and thermal resistance, along with improved strength. This section explores current material trends in LC, detailing key compositions, properties, and applications—summarized in Fig. 7.4.

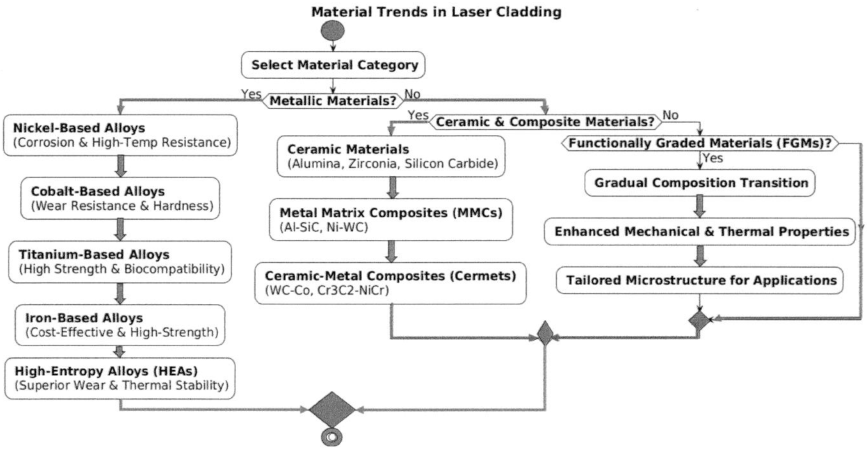

Figure 7.4: Material system used in the LC.

7.3.1 Metallic Materials

Metallic materials are the primary feedstock in laser cladding (LC) due to their strong mechanical properties, substrate compatibility, and broad application range. This section covers widely used alloys and highlights emerging trends in metallic feedstock design.

- **Nickel-based alloys**

Nickel-based alloys are widely favored in laser cladding (LC) for their excellent corrosion resistance, thermal stability, and wear performance. Common variants like Inconel 625, Inconel 718, and Hastelloy are extensively applied in aerospace, energy, and chemical industries. Recent progress has focused on enhancing their mechanical strength and thermal resilience for high-temperature environments. (Zhang, 2024) reported that NiCrFeWBSiMo-CeO_2 coatings deposited via LC exhibited excellent surface quality with uniform, defect-free deposition, as shown in Fig. 7.5(a), confirming optimized process parameters. The microhardness profile in Fig. 7.5(b) shows an average hardness of ~575 HV, indicating significantly enhanced mechanical properties over conventional coatings.

Figure 7.5(c) illustrates the tribological performance of the Ni-based alloy coating, which shows a coefficient of friction (COF) of 0.39 and a wear rate of 1.1 mg. The COF curve displays two phases: an initial rise due to two-body friction with the Al_2O_3 counterface, followed by stabilization—signaling the formation of a tribo-layer that governs steady-state wear behavior. Figure 7.5(d-e) reveals narrow, smooth wear tracks, indicating excellent wear resistance of the Ni-based coating. Observed features like furrows, debris, and

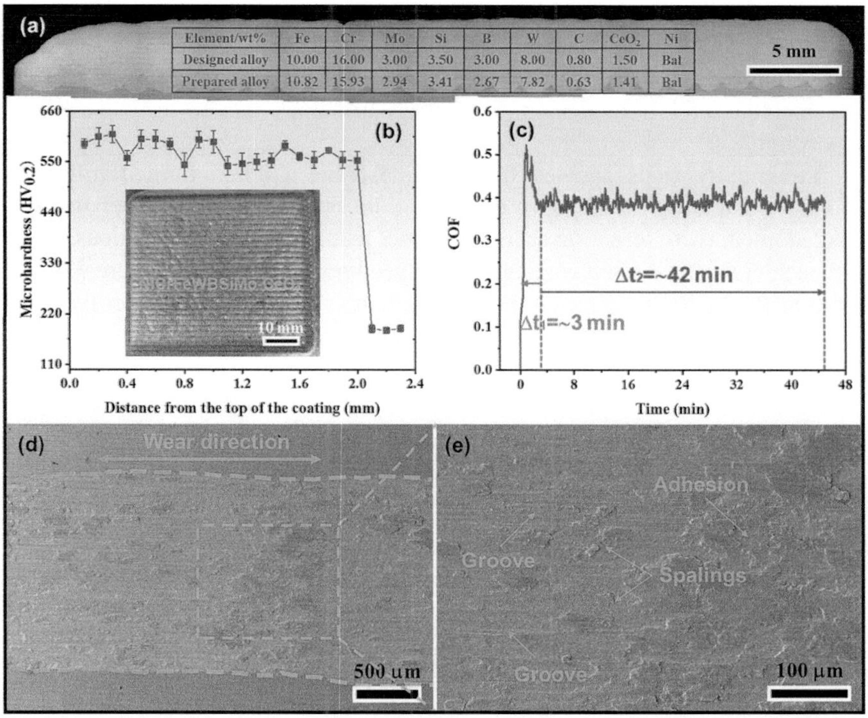

Figure 7.5: (a) Wide-field macroscopic view of the coated surface; (b) Microhardness distribution profile from the coating to the substrate alongside the coating morphology; (c) Variation of the friction coefficient over time; (d-e) SEM images illustrating the wear (Zhang, 2024).

abrasive chips point to mixed adhesive–abrasive wear. This superior performance is attributed to the coating's refined microstructure, which enhances its resistance to sliding-induced degradation.

- **Cobalt-based alloys**

Cobalt-based alloys—particularly Stellite 6 and Stellite 21—are widely used in laser cladding (LC) due to their outstanding wear resistance, hardness, and thermal fatigue strength. These properties make them ideal for demanding applications like turbine blades, engine parts, and cutting tools. Recent research focuses on enhancing their toughness and oxidation resistance for improved performance in extreme conditions. (Zhao et al., 2024) investigated laser cladding (LC) of Stellite 21 alloy blended with 1–10% manganese powder to improve material properties. Stellite 21 (spherical, Höganäs AB) and Mn powder (blocky, 45–105 μm, Changsha Tijo) were mixed via planetary ball milling at 200 rpm for 3 hours. LC was performed on 24CrNiMo low-alloy steel, with powder and substrate morphologies shown in Fig. 7.6(a-c).

The LDED process used an IPG YLS-12000 system with a 12 kW fiber laser, delivered through optics and focused to a 4.5 mm spot via a 15 mm cladding head. High-purity argon (99.9999%) was supplied for powder transport (4 L/h) and shielding (25 L/h) to prevent oxidation. A 40% track overlap ensured uniform deposition.

Preliminary trials showed that *PFR*, *Ss*, and *Lp* had limited influence on cladding quality. To improve deposition, an orthogonal multi-layer strategy was adopted, with each layer oriented perpendicular to the previous. This approach yielded a ~4 mm coating with reduced residual stresses. Figure 7.6(d-f) illustrate the LDED setup, while Fig. 7.6(g) depicts the continuous layer-by-layer process, which enhanced structural uniformity and mechanical stability.

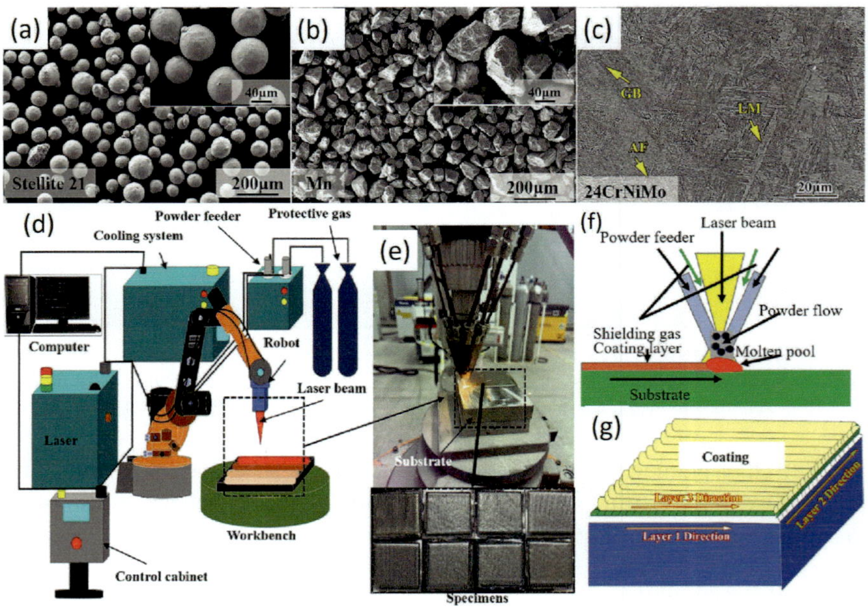

Figure 7.6: SEM images showing the morphological characteristics of different materials: (a) Stellite 21 alloy, (b) Manganese powder, and (c) 24CrNiMo (d) Deposition equipment, (e) Step-by-step deposition process, (f) Schematic illustration of the deposition method, (g) Layout of the coating formation process (Zhao et al., 2024).

- **Titanium-based alloys**

Titanium-based alloys, such as Ti-6Al-4V and Ti-6Al-7Nb, are widely used in laser cladding (LC) for aerospace, biomedical, and marine applications due to their high strength-to-weight ratio, corrosion resistance, and biocompatibility. Recent studies emphasize microstructure and property optimization through process tuning and post-treatment techniques.

(Hu et al., 2024) employed a sequential laser cladding approach, first depositing Ni22AA alloy on Q235 steel to form a Ni interlayer, followed by TC4 powder to create the final coating. Using a 40% track overlap and an "n-shaped" scanning path (Fig. 7.7a), uniform multi-track deposition was achieved. The Ni interlayer exhibited good surface integrity (Fig. 7.7c), promoting strong TC4 adhesion. As shown in Fig. 7.7(d, g-h), specimens SNT-1, SNT-4, and

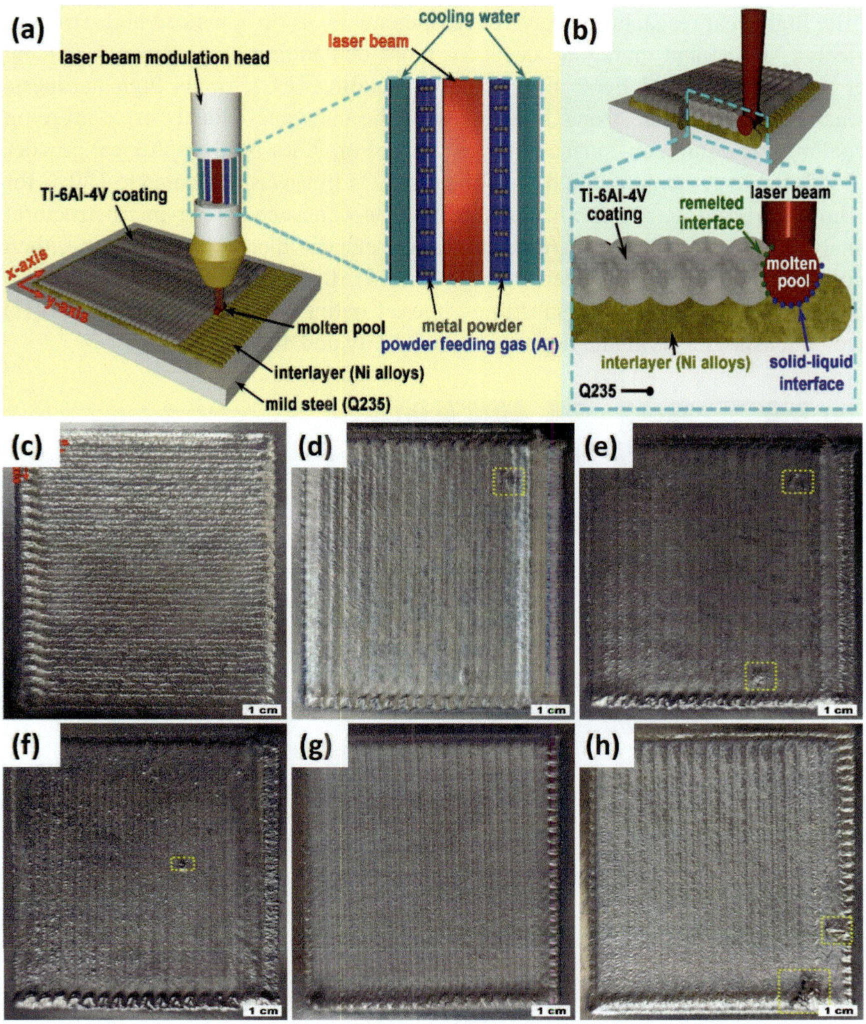

Figure 7.7: Schematic representation of the TC4 coating with a nickel interlayer: (a) LC procedure, (b) Cross-sectional metallographic structure; Macrostructural characteristics of TC_4 coatings fabricated using various cladding techniques: (c) Ni interlayer, (d) SNT-1, (e) SNT-2, (f) SNT-3, (g) SNT-4, (h) SNT-5 (Hu et al., 2024).

SNT-5 had smooth surfaces with minor defects, while SNT-2 and SNT-3 initially showed rough morphology (Fig. 7.7e-f) that improved during deposition.

- **Iron-based alloys**

Iron-based alloys are commonly used in laser cladding (LC) due to their affordability, availability, and solid mechanical performance. Stainless steels like 316L and 420 provide corrosion resistance, while tool steels such as H13 and D2 offer high wear resistance. Recent developments focus on advanced high-strength steels (AHSS) and maraging steels for more demanding environments. (Zhang et al., 2025) applied Fe60 alloy in laser cladding (LC) for its high hardness, wear resistance, and strong bonding with the substrate—ideal for reinforcing coal mining equipment surfaces. As shown in Fig. 7.8(a-b), the spherical powder (8–55 μm, avg. diameter: 32.67 μm, mean: 27.25 μm) was preheated to 120°C for 2 hours to improve flowability and deposition consistency. For microstructural analysis, coated samples were sectioned via electric discharge machining, then polished and etched. As seen in Fig. 7.8(c-d), the cross-section reveals clear coating, heat-affected, and substrate zones. Coating width (W) and height (H) were measured to evaluate deposition uniformity and adhesion.

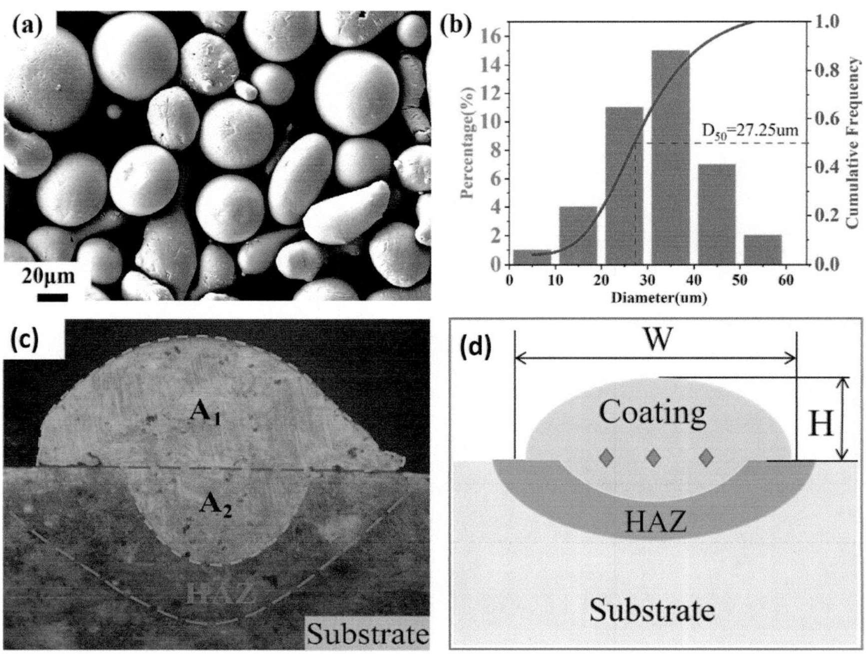

Figure 7.8: Fe60 alloy powder characterization: (a) SEM micrograph showing the powder morphology, (b) Particle size distribution analysis; (c-d) Cross-sectional view of a single-layer coating and its interface with the substrate (Zhang et al., 2025).

- **High-entropy alloys (HEAs)**

High-entropy alloys (HEAs), composed of multiple principal elements in near-equimolar ratios, offer exceptional strength, wear resistance, and thermal stability (Zhang Qinghua et al., 2025; Liu et al., 2025; Zheng et al., 2025). These properties make them promising candidates for laser cladding (LC) in demanding environments. Common HEAs explored in LC include FeCoCrNiMo, AlCoCrFeNi, and CoCrFeNiMn variants (Feng et al., 2024).

7.3.2 Ceramic and Composite Materials

Ceramic and composite materials are increasingly used in laser cladding (LC) to enhance coating hardness, abrasion resistance, and thermal stability.

- **Ceramic materials**

In laser cladding (LC), ceramics like alumina (Al_2O_3), zirconia (ZrO_2), and silicon carbide (SiC) are used to produce hard, wear-resistant coatings ideal for tools, wear plates, and thermal barriers. However, their brittleness can cause cracks and bonding issues. To address this, recent efforts focus on ceramic-metal composites and functionally graded materials (FGMs) to improve toughness and interfacial integrity. (Chen et al., 2019) employed nanostructured TiC ceramic powder (50 nm, 99.9% purity) as reinforcement in LC. SEM and TEM images in Fig. 7.9(a–b) confirm its uniform morphology and distribution. To ensure homogeneous dispersion, TiC was suspended in polyvinyl alcohol solution,

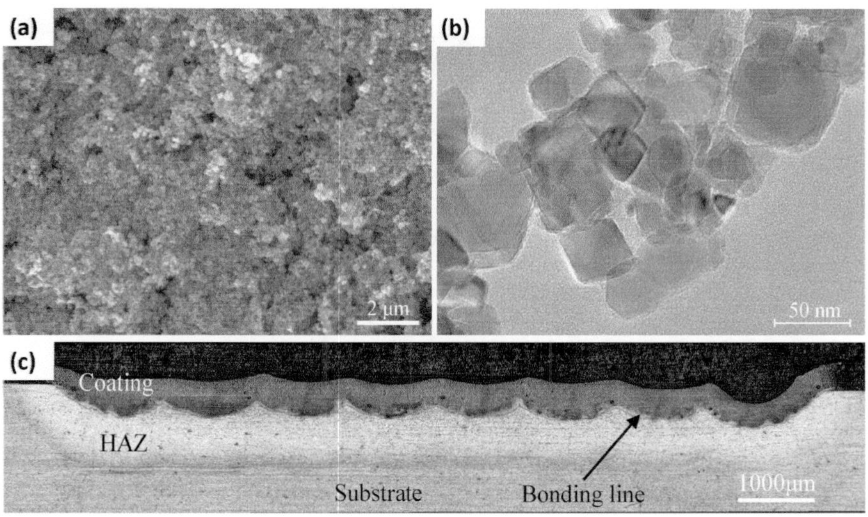

Figure 7.9: TiC nanoparticles utilized in this study: (a) SEM micrograph, (b) TEM micrograph, (c) Cross-sectional image of a representative sample produced through LC with TiC nanoparticle reinforcement (Chen et al., 2019).

stirred, and spread onto the substrate surface, then air-dried to form a pre-positioned layer. The resulting multi-track coating's cross-section is shown in Fig. 7.9(c) via digital microscopy.

- **Metal matrix composites**

Metal matrix composites (MMCs) combine metal toughness with ceramic hardness, making them well-suited for laser cladding (LC). Common MMCs include Al–SiC and Ni–WC, offering excellent wear resistance and durability. Recent advances focus on nano-reinforcements to further improve mechanical properties and performance. (Zhao et al., 2022) used Fe/WC tubular cored wire for laser wire cladding. As shown in Fig. 7.10(a), a U-shaped pure iron strip was filled with ceramic and alloying powders, sealed, and drawn to a 1.6 mm diameter. The cross-section confirms uniform powder distribution. Spherical WC particles (~100 μm) served as reinforcements due to their high thermal stability. XRD analysis (Fig. 7.10b) confirmed WC and W_2C phases, while SEM (Fig. 7.10c) highlighted their eutectoid structure, enhancing coating hardness and wear resistance. Cladding was performed using a 2 kW CW Nd:YAG laser (1.064 μm), focused to 4 mm with a 300 mm lens. A wire feeder ensured steady material delivery (Fig. 7.10d).

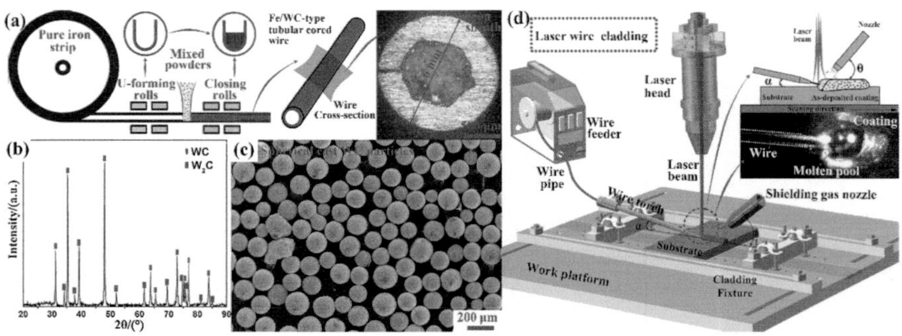

Figure 7.10: Material preparation and experimental setup: (a) Schematic illustration of the WC-Fe tubular cored wire manufacturing process, (b) XRD spectra of tungsten carbide phases, (c) SEM micrograph of spherical tungsten carbide particles, (d) Experimental configuration for laser wire deposition (Zhao et al., 2022).

- **Ceramic-metal composites**

Ceramic-metal composites—combine the hardness of ceramics with the toughness of metals, offering a balanced solution for laser cladding (LC). They're commonly applied in cutting tools, wear-resistant layers, and thermal barrier coatings. Popular cermets like WC–Co and Cr_3C_2–NiCr are widely used, with recent work focusing on refining their composition and microstructure to enhance durability and performance. (Shen et al. 2022) laser-cladded three NiCrSiBC–WC composite powders—Powder 1 (40% WC), Powder 2 (50% WC), and Powder 3 (60% WC)—onto medium-carbon steel (0.45% C), cut to 40 × 60 × 8 mm. SEM images in Fig. 7.11(a-c) show WC as white blocky particles and NiCrSiBC as spherical ones. Due to LC's rapid heating–cooling cycles, notable phase transformations occur. Figure 7.11(d) displays uniform cladded surfaces with distinct tracks and no impurities, while post-polishing morphology in Fig. 7.11(e) reveals a smoother finish for detailed analysis.

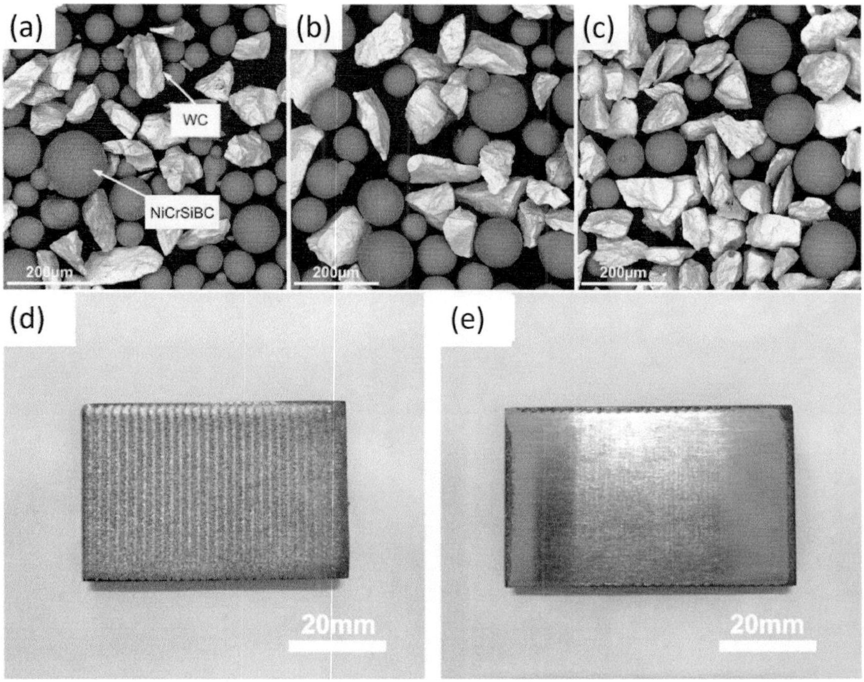

Figure 7.11: Powder morphology analysis: (a) WC with 40% weight fraction, (b) WC with 50% weight fraction, (c) WC with 60% weight fraction; (d-e) Surface morphology of the laser-cladded coating at the macroscopic scale (Shen et al., 2022).

7.3.3 Functionally Graded Materials (FGMs)

Functionally graded materials (FGMs) feature gradual changes in composition and properties, enabling tailored performance across their structure. Laser cladding (LC) is ideal for fabricating FGMs due to its precise material placement. For example, coatings can transition from metallic at the substrate to ceramic at the surface, balancing strength and hardness. FGMs are widely used in wear-resistant and thermal barrier coatings, as well as biomedical implants designed to mimic the mechanical properties of natural bone. (Ferreira et al., 2025) fabricated a functionally graded material (FGM) using MetcoClad 625

(M625) and Diamalloy 4006 (D4006) powders, delivered via dual-hopper feed through a coaxial nozzle. Laser cladding was performed with a 2.5 mm spot and 0.2 mm Z-step, under 99.99% pure argon at 5.5 L/min to avoid oxidation. M625 (gas-atomized, 45–90 μm) had spherical morphology, while D4006 (water-atomized, 11–53 μm) showed quasi-spheroidal shape (Fig. 12a-b). The resulting FGM exhibited vertically aligned columnar γ-dendrites with refined secondary arms and narrower spacing due to rapid cooling and grain-refining elements (Fig. 12c).

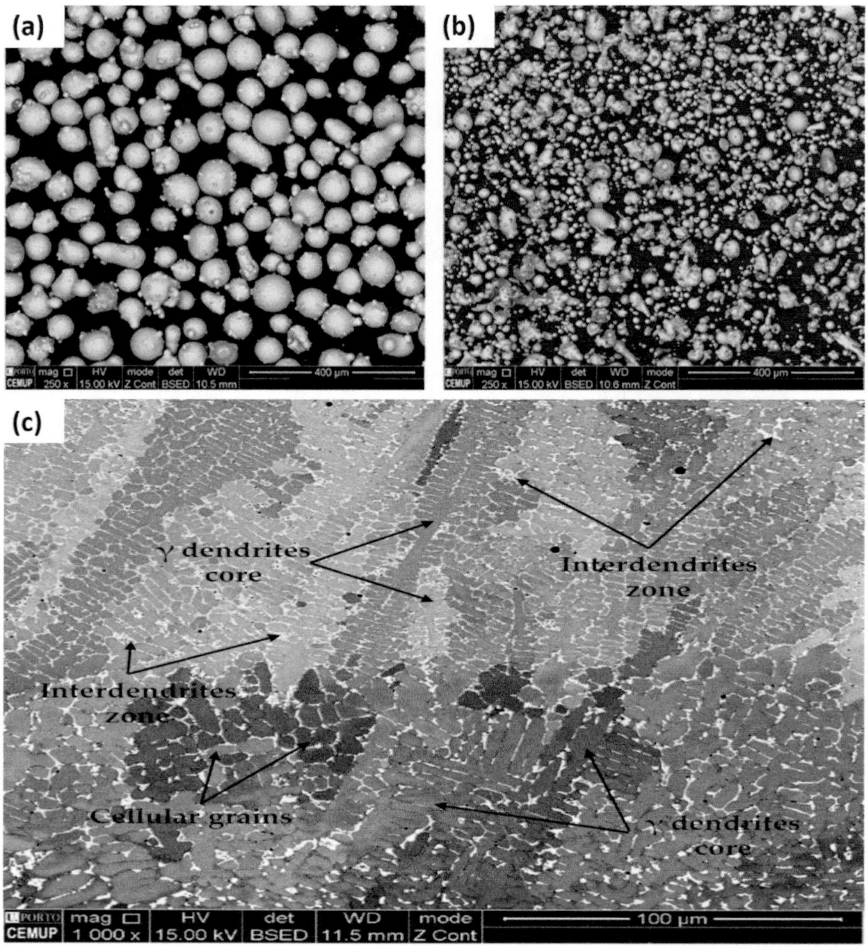

Figure 7.12: Morphological analysis of powders: (a) MetcoClad 625 (M625), (b) NiCrWMo (D4006) (c) SEM micrograph illustrating the microstructural features of FGM, focusing on the transition zone between M625–25% D4006 and M625–50% D4006 (Ferreira et al., 2025).

7.4 Material Characterization and Quality Assurance

The performance of laser-clad components depends heavily on the quality of the deposited material. This section outlines key techniques for evaluating material properties and ensuring quality throughout the laser cladding (LC) process.

7.4.1 Microstructural Evaluation

Microstructural evaluation is key to understanding the mechanical performance of laser-clad materials. Techniques like SEM, TEM, and XRD are commonly used to examine microstructure, phase distribution, and crystallographic features of the clad layer.

7.4.2 Assessment of Mechanical Properties

Material suitability in laser cladding is largely determined by mechanical properties—such as wear resistance, tensile strength, and microhardness. Standard evaluations like hardness tests, wear analysis, and tensile strength measurements ensure the deposited layers meet application-specific performance benchmarks.

7.4.3 Non-Destructive Evaluation (NDE)

Non-destructive evaluation (NDE) techniques ensure the reliability of laser-clad layers by detecting defects like inclusions, porosity, cracks, and lack of fusion without damaging the component. Common methods include Barkhausen noise, X-ray radiography, ultrasonic testing, and eddy current inspection.

7.5 Applications of LC

Laser cladding (LC) enables high-quality coatings, component repair, and complex geometry fabrication—making it essential in modern manufacturing. The following sections highlight its major applications across aerospace, automotive, energy, and biomedical sectors (see Fig. 7.13 for an overview).

7.5.1 Industrial Applications

Laser cladding (LC) is widely employed in industrial environments for surface enhancement, part repair, and additive manufacturing. Key applications include:

- **Repair and Restoration of High-Value Components**

Laser cladding (LC) effectively restores worn or damaged components—such as turbine blades, molds, dies, and shafts—extending their service life and reducing costly replacements. In the oil and gas industry, it's widely used to refurbish valves and drill bits, ensuring reliable performance under harsh conditions.

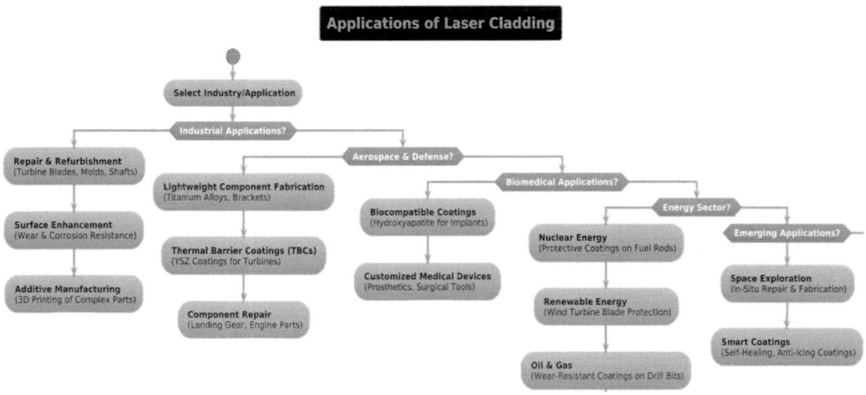

Figure 7.13: Applications of the LC in different sectors.

- **Surface Enhancement for Wear and Corrosion Resistance**

Laser cladding (LC) is widely used for applying wear- and corrosion-resistant coatings that boost thermal stability. In mining, such coatings extend the life of conveyor parts and excavator teeth, reducing maintenance and downtime.

- **Additive Manufacturing and 3D Printing**

Laser cladding (LC) enables the fabrication of complex, customized parts—crucial for aerospace and medical sectors. It supports designs like turbine blades with internal cooling channels, boosting performance and efficiency.

7.5.2 Aerospace and Defense

The aerospace industry has led in adopting laser cladding (LC) to boost the performance of critical parts. Titanium alloys, commonly used in this sector, benefit from LC through enhanced wear resistance and high-temperature oxidation protection.

- **Lightweight Component Fabrication**

Laser cladding (LC) enables the fabrication of lightweight, intricate aerospace parts—such as titanium alloy brackets and fittings—reducing overall weight and enhancing fuel efficiency. Its precision and material control make LC ideal for high-performance component manufacturing.

- **Component Repair and Refurbishment**

Laser cladding (LC) helps the aerospace industry cut maintenance costs and downtime by refurbishing key components—like turbine blades, landing gear, and engine parts—while maintaining original performance standards.

7.5.3 Biomedical Applications

Laser cladding (LC) enhances biocompatibility and corrosion resistance, making it ideal for manufacturing and coating medical implants and surgical tools.

- **Biocompatible Coatings for Implants**

Laser cladding (LC) enhances medical implants by enabling biocompatible coatings that improve integration and reduce rejection risks. For instance, HA-coated dental and orthopedic implants promote bone growth, while laser-clad titanium alloys offer strong mechanical performance and high biocompatibility.

7.5.4 Energy Sector

Laser cladding (LC) supports the energy sector by enabling surface enhancement, component repair, and additive manufacturing across nuclear, renewable, and fossil fuel systems. In nuclear power, it protects reactor parts—like fuel and control rods—from corrosion, wear, and radiation, while also aiding in repair to reduce downtime and improve safety. In renewables, LC coatings enhance durability and lifespan of wind turbine blades and solar panel components by preventing wear and corrosion.

7.5.5 Emerging Applications

Laser cladding (LC) is rapidly evolving with emerging applications across diverse fields. In space exploration, LC shows promise for *in situ* fabrication and repair of tools and spacecraft components due to its precision and material efficiency. It's also being used to create advanced coatings with anti-fouling, anti-icing, and self-healing properties. Embedded sensors in laser-clad layers enable real-time health monitoring, enhancing safety and reliability. Similarly, hydrophobic laser-clad coatings can enhance safety and performance by preventing ice accumulation on aircraft wings.

7.6 Case Study: Multi-Track Deposition of FeCoCrNiMo HEA on Steel Substrate

FeCoCrNiMo high-entropy alloy (HEA) coating was laser-clad onto steel substrate to improve surface properties. The process began with single-track trials to optimize parameters, followed by a multi-track, single-layer deposition using the best settings (see Fig. 7.14). Microstructural and hardness analyses confirmed a high-quality, defect-free coating with superior performance over the base material.

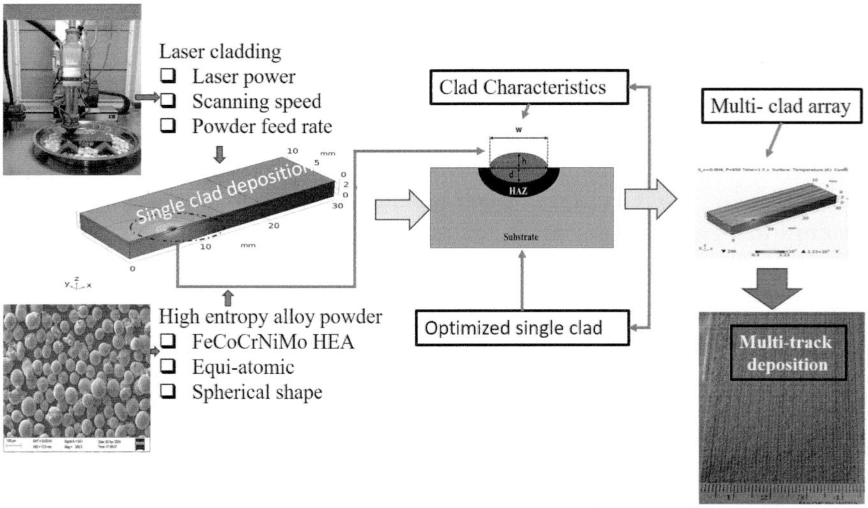

Figure 7.14: Schematic representation of the multi-track single layer deposition of FeCoCrNiMo HEA using LC.

To ensure successful deposition, single-track trials were first conducted by varying laser power (Lp), scanning speed (Ss), and powder feed rate (PFR). These trials identified optimal parameters—1530 W Lp, 3 mm/s Ss, and 11.556 g/min PFR—yielding uniform tracks with minimal defects and strong metallurgical bonding. A subsequent multi-track, single-layer deposition using these settings achieved consistent, defect-free coatings across overlapping tracks. Post-deposition, optical microscopy Fig. (7.15) revealed a refined dendritic microstructure with uniform elemental dispersion—typical of HEAs. The HEA-clad layer showed finer grains than steel substrate, attributed to rapid solidification and thermal gradients during laser cladding.

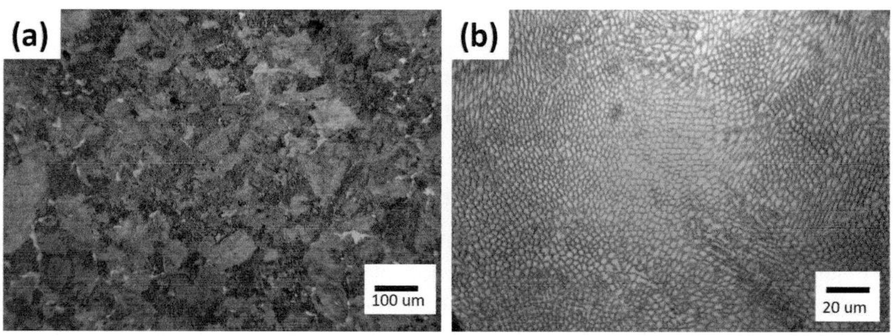

Figure 7.15: Microstructure of (a) Steel substrate (b) HEA clad.

Figure 7.16 shows Vickers hardness results for the HEA-cladded layer, revealing a significant increase over the base steel substrate. The smooth hardness gradient across the interface indicates strong bonding and consistent properties. This enhancement suggests improved wear resistance, making the coating suitable for high-stress, wear-intensive applications.

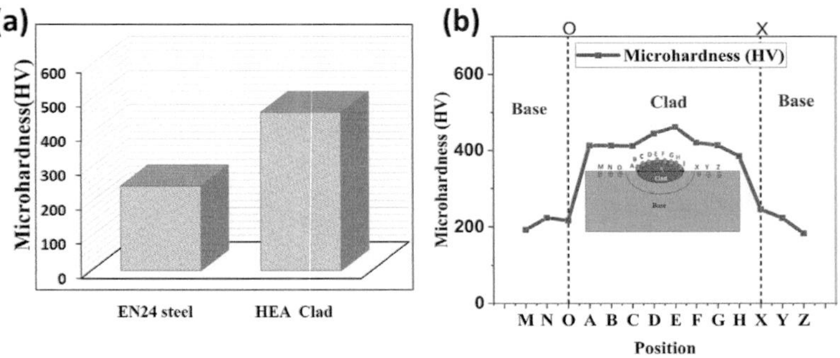

Figure 7.16: Microhardness variation of the steel substrate and HEA clad.

7.7 Challenges and Limitations of LC

Laser cladding (LC) faces several limitations that often hinder its full potential. Rapid thermal cycles can induce residual stresses, leading to distortion, delamination, or cracking in both the substrate and clad layer (Diwakar et al., 2025). Common issues like porosity and incomplete fusion arise from poor shielding gas flow, inadequate powder melting, or fast solidification. Additionally, LC's high sensitivity to parameter fluctuations makes reproducibility challenging. The process also demands significant capital investment in lasers, feeders, and motion systems—alongside high operating costs, particularly in energy consumption. To meet green manufacturing goals, challenges such as material waste and gas emissions must be addressed. Promising solutions include eco-friendly feedstocks, advanced simulations for defect prediction, and optimized process controls.

7.8 Future Perspectives and Innovations

Laser cladding (LC) continues to evolve with advancements in materials and processing. High-entropy alloys (HEAs), comprising five or more principal elements in equimolar or near-equimolar ratios, offer enhanced hardness, strength, and corrosion resistance due to their solid-solution BCC or FCC phases. For instance, AlCoCrFeNi HEA coatings with a BCC/B2 structure exhibit notably high microhardness (Chen et al., 2024). Mo additions can reduce corrosion resistance, but CoCrFeNiMo HEAs show excellent saltwater corrosion performance due to protective oxide layers (Qi Zhang et al., 2023). Incorporating ceramics like TiC, WC, and NbC further boosts wear resistance and hardness (Tang et al., 2024; Han and Fu, 2024). Enhanced high-speed laser cladding (EHLA) offers high efficiency, improved surface accuracy, and a reduced heat-affected zone, making it well-suited for complex HEA coatings. Its superior powder utilization surpasses that of conventional LC (Zhou et al., 2024). Hybrid LC—combining LC with methods like WAAM—addresses traditional limitations, boosting efficiency, flexibility, and mechanical performance. This approach has successfully produced high-toughness stainless steel parts (Chen et al., 2024). Multilayer cladding further enhances hardness, wear, and corrosion resistance for extreme-condition applications.

7.9 Conclusion

LC Laser cladding (LC) was introduced to enhance coating performance with minimal added weight, surpassing conventional surface techniques. This chapter examines key LC elements, including microstructural evolution and changes in mechanical and tribological properties. Advances in alloys, FGMs, and composite coatings now enable high-performance surfaces with superior hardness, corrosion, and wear resistance. Notably, Ni- and Fe-based alloys reinforced with WC and TiC ceramics have shown exceptional results in improving hardness and durability. Microstructural transitions from columnar to equiaxed structures significantly enhance coating toughness and load-bearing capacity due to dendritic growth. Advanced LC techniques—like direct laser and FGM deposition—enable precise control over these transitions, improving material properties. SEM and TEM analyses confirm low-porosity, defect-free coatings with optimal grain structures.Despite notable progress, challenges remain in boosting deposition efficiency, minimizing thermal distortion, and adapting parameters across material systems. Future research should leverage AI and machine learning for real-time process optimization and property enhancement.

References

Chen, Tao, Weining Wu, Wenpeng Li and Defu Liu. 2019. Laser cladding of nanoparticle TiC ceramic powder: Effects of process parameters on the quality characteristics of the coatings and its prediction model. *Optics and Laser Technology*, 116(March): 345–55. https://doi.org/10.1016/j.optlastec.2019.03.048.

Chen, Zheng, Zhenghua Guo, Yida Zeng, Zhiyong Li, Yan Wang et al. 2024. Study on microstructure and properties of AlCoCrFeNi high-entropy alloys by extreme high-speed laser cladding. *Coatings*, 14(11): 1–13. https://doi.org/10.3390/coatings14111394.

Diwakar, Vikas, Ashwani Sharma, Mohd Zaheer Khan Yusufzai and Meghanshu Vashista. 2024. Direct energy deposited cladded material characterization: exploring barkhausen noise and thermal field with parametric simulation. *Russian Journal of Nondestructive Testing*, 60(2): 223–37. https://doi.org/10.1134/S1061830923600806.

Diwakar, Vikas, Ashwani Sharma, Mohd Zaheer Khan Yusufzai and Meghanshu Vashista. 2025. Micro-magnetic non-destructive characterization of multi-clad array coating of FeCoCrNiMo high entropy alloy in laser cladding. *NDT and E. International*, 155(April): 103428. https://doi.org/10.1016/j.ndteint.2025.103428.

Feng, Meiyan, Tianxiang Lin, Guofu Lian, Changrong Chen, Xu Huang et al. 2024. Effects of Nb content on the microstructure and properties of CoCrFeMnNiNbx high-entropy alloy coatings by laser cladding. *Journal of Materials Research and Technology*, 28(September 2023): 3835–48. https://doi.org/10.1016/j.jmrt.2024.01.002.

Ferreira, A., Ana R. Reis and Sousa, P. 2025. Materials characterization inconel 625/NiCrWMo functionally graded materials manufactured by direct laser deposition, 222(February). https://doi.org/10.1016/j.matchar.2025.114862.

Han, Yantao and Hanguang Fu. 2024. Improvement of high temperature wear resistance of laser-cladding high-entropy alloy coatings: A review. *Metals*, 14(9). https://doi.org/10.3390/met14091065.

Hao, Liang, Wanlin Wang, Jie Zeng, Min Song, Shuai Chang et al. 2023. Effect of scanning speed and laser power on formability, microstructure, and quality of 316L stainless steel prepared by selective laser melting. *Journal of Materials Research and Technology*, 25: 3189–99. https://doi.org/10.1016/j.jmrt.2023.06.144.

Hu, Kangkai, Yuxin Tian, Xuzhou Jiang, Hongying Yu, Dongbai Sun et al. 2024. Microstructure regulation and performance of titanium alloy coating with ni interlayer on the surface of mild steel by laser cladding. *Surface and Coatings Technology*, 487(March): 130939. https://doi.org/10.1016/j.surfcoat.2024.130939.

Imran, Muhammad Mu'az, Azam Che Idris, Liyanage Chandratilak De Silva, Yun Bae Kim, Pg Emeroylariffion Abas et al. 2024. Advancements in 3D printing: Directed energy deposition techniques, defect analysis, and quality monitoring. *Technologies*, 12(6). https://doi.org/10.3390/technologies12060086.

Jin, Xueying, Tingwen Gu, Ye Zhan, Dongming Qu, Guanyu Chen et al. 2024. A review: Experimental and applied research of double-pulse laser-induced breakdown spectroscopy. *Applied Spectroscopy Reviews*, 59(9): 1278–1321. https://doi.org/10.1080/05704928.2024.2320146.

Li, Zuo, Shang Sui, Xu Ma, Hua Tan, Chongliang Zhong et al. 2022. High deposition rate powder- and wire-based laser directed energy deposition of metallic materials: A review. *International Journal of Machine Tools and Manufacture*, 181(June). https://doi.org/10.1016/j.ijmachtools.2022.103942.

Liu, Hao, Baochen Wang, Peijian Chen, Dali Li, Jingbin Hao et al. 2025. Hierarchical heterogeneous microstructure for enhanced wear resistance of CoCrFeMnNi high-entropy alloy coatings via *in situ* rolling assisted laser cladding. *Journal of Manufacturing Processes*, 141(February): 105–20. https://doi.org/10.1016/j.jmapro.2025.02.066.

Niu, Fangyong, Haopeng Xing, Fangyan Kan, Guangyi Ma, Xiong Sun et al. 2024. Remanufacturing of 3D-damaged components based on LDED technology. *Journal of Materials Engineering and Performance*, 33(14): 7378–92. https://doi.org/10.1007/s11665-023-08464-6.

Rong, Chuman, Xia Liu, Yaqi Wang, Shangfei Qiao, Xiaojian Xu et al. 2025. Study on the formation mechanism of HAZ microcracks in FB2 heat-resistant steel by laser cladding. *Materials Characterization*, 221(December 2024): 114750. https://doi.org/10.1016/j.matchar.2025.114750.

Shen, Xuehui, Xiangping He, Liying Gao, Guosheng Su, Chonghai Xu et al. 2022. Study on crack behavior of laser cladding ceramic-metal composite coating with high content of WC. *Ceramics International*, 48(12): 17460–70. https://doi.org/10.1016/j.ceramint.2022.03.010.

Tang, Junli, Kaiming Wang and Hanguang Fu. 2024. Laser cladding *in situ* carbide-reinforced iron-based alloy coating: A review. *Metals*, 14(12): 1–22. https://doi.org/10.3390/met14121419.

Xiang, Dingding, Yusheng Liu, Tianbiao Yu, Di Wang, Xiaoxin Leng et al. 2024. Review on wear resistance of laser cladding high-entropy alloy coatings. *Journal of Materials Research and Technology*, 28(August 2023): 911–34. https://doi.org/10.1016/j.jmrt.2023.11.138.

Xie, Xiao Ming, Xiu Bo Liu, Bo Ming He, Fei Zhi Zhang, Ji Xiang Liang et al. 2025. Laser cladding Ni60-SiC/Ti3SiC2 self-lubricating composite coatings on IN718 alloy: Wear mechanisms and oxidation behaviors. *Wear*, 560-561(October 2024): 205611. https://doi.org/10.1016/j.wear.2024.205611.

Yamaguchi, Takuto, Keigo Tanaka and Hideki Hagino. 2023. Porosity reduction in WC-12Co laser cladding by aluminum addition. *International Journal of Refractory Metals and Hard Materials*, 110(June 2022): 106020. https://doi.org/10.1016/j.ijrmhm.2022.106020.

Yamaguchi, Takuto, Keigo Tanaka and Takeshi Suyama. 2025. *In-situ* observation of CO gas bubble behavior in molten pool during direct laser deposition of WC-Co cermet material. *Optics and Laser Technology*, 180(August 2024). https://doi.org/10.1016/j.optlastec.2024.111561.

Yuan, Junfeng, Heran Geng and Marco Alfano. 2024. Multi-response optimization of M2 steel coatings deposited by Co-axial laser cladding on A2 steel tool surfaces. *Journal of Materials Research and Technology*, 29(November 2023): 1102–17. https://doi.org/10.1016/j.jmrt.2024.01.087.

Zhang, B. 2024. Microstructure and wear resistance of laser cladding a novel nickel-based alloy coating. *Materials Letters*, 357(September 2023): 135695. https://doi.org/10.1016/j.matlet.2023.135695.

Zhang, Qi, Meiyan Li, Qin Wang, Fuhao Qi, Mengkai Kong et al. 2023. Investigation of the microstructure and properties of CoCrFeNiMo high-entropy alloy coating prepared through high-speed laser cladding. *Coatings*, 13(7). https://doi.org/10.3390/coatings13071211.

Zhang, Qinghua, Hongxi Liu, Ling Zhou, Jiazhu Liang, Chen Yang et al. 2025. Microstructure and wear-resistance of AlNbTiVB x high entropy alloy coatings prepared by laser cladding on Ti6Al4V substrate. *Materials Today Communications*, 44(November 2024): 112158. https://doi.org/10.1016/j.mtcomm.2025.112158.

Zhang, Wei, Xianhe Shang, Minglei Hu, Xing He, Bing Yang, Kunjie Dai, Xiaoqing Ni et al. 2023. Microstructure and corrosion-wear behaviors for laser cladding repaired martensitic stainless steels using Co-based and Ni-based powders. *Materials Today Communications*, 35(March): 106287. https://doi.org/10.1016/j.mtcomm.2023.106287.

Zhang, Yi, Peikang Bai, Zhonghua Li and Jie Zhang. 2025. Rapid optimization of iron-based alloy laser cladding process based on orthogonal experiment and machine learning for Q345. *Optics and Laser Technology*, 182(PA): 112086. https://doi.org/10.1016/j.optlastec.2024.112086.

Zhao, Shengbin, Sai Xu, Yiming Huang and Lijun Yang. 2021. Laser hot-wire cladding of Ni/WC composite coatings with a tubular cored wire. *Journal of Materials Processing Technology*, 298(June): 117273. https://doi.org/10.1016/j.jmatprotec.2021.117273.

Zhao, Shengbin, Sai Xu, Lijun Yang and Yiming Huang. 2022. WC-Fe metal-matrix composite coatings fabricated by laser wire cladding. *Journal of Materials Processing Technology*, 301(October 2021): 117438. https://doi.org/10.1016/j.jmatprotec.2021.117438.

Zhao, Yongsheng, Ying Wu, Yan Liu, Zhenlin Zhang, Xiaoli Che et al. 2024. Effect of Mn alloying on the microstructure and micro-shear properties of laser direct energy deposition Stellite21 cobalt-based coating and its process research. *Surface and Coatings Technology*, 483(April): 130786. https://doi.org/10.1016/j.surfcoat.2024.130786.

Zheng, K.X.D.T., Yu, J.L., Liu, C.L., Wu, S., Zhang, C.H. et al. 2025. Laser cladding of FeCoCrNiTi high-entropy alloy coatings to modulate the microstructure and enhance the tribo-corrosion behavior on 304 stainless steel. *Surface and Coatings Technology*, 505(April). https://doi.org/10.1016/j.surfcoat.2025.132114.

Zhou, Li, Guozheng Ma, Haichao Zhao, Honglin Mou, Jianfeng Xu et al. 2024. Research status and prospect of extreme high-speed laser cladding technology. *Optics and Laser Technology*, 168(April 2023): 109800. https://doi.org/10.1016/j.optlastec.2023.109800.

CHAPTER

8 3D Printing and its Impact on Composite Materials Development

Priya Singh,[1] *Ravi Prakash Ojha*[2]
and *Santosh Kumar*[1,*]

3D printing has transformed materials research, especially in composites, by enabling complex, tailored structures unattainable through traditional methods. This chapter reviews key 3D printing techniques and their applications in composites—spanning polymers, metals, and ceramics. It explores how additive manufacturing enhances design flexibility, reduces weight, and improves performance across aerospace, biomedical, and automotive sectors. Challenges like scale-up, material limitations, and regulations are also addressed. The chapter concludes with a forward-looking view of 3D printing's role in sustainable, next-generation composite manufacturing.

8.1 Introduction

3D printing, or additive manufacturing, creates three-dimensional objects by layering material based on digital designs. Unlike traditional subtractive methods, it enables the fabrication of intricate, customized parts without the need for molds or tooling (Kulkarni and Sharma, 2024). Its versatility spans a broad material range—from polymers and metals to food ingredients—allowing production from large structures to fine or edible items (Barnatt, 2013; Choonara et al., 2016; Singh and Gupta, 2024).

[1] Department of Chemistry, Harcourt Butler Technical University Kanpur, Kanpur India.
[2] Department of Chemistry, Shri Lal Bahadur Shastri Degree College, Gonda, India.
* Corresponding author: santoshk@hbtu.ac.in

DOI: 10.1201/9781003607359-8

3D printing spans diverse disciplines—from prototyping and manufacturing to medicine, construction, and art. It enables rapid prototyping, cutting time and cost in product development. Additive manufacturing offers design freedom and supports low-volume, customized production across sectors like consumer goods, aerospace, and automotive (Joshi and Sheikh, 2015; Shahrubudin et al., 2019). Incorporating advanced polymer composites enhances mechanical, thermal, and electrical performance for demanding applications (Kulkarni and Sharma, 2024; Ma and Zhang, 2024). Additionally, metal 3D printing techniques like electron beam melting (EBM) and laser powder bed fusion (LPBF) enable complex, lightweight structures (Peels, 2024).

3D printing is transforming healthcare by enabling bioprinting of tissues and the production of patient-specific implants, prosthetics, and dental devices—driving personalized medicine forward (Kulkarni and Sharma, 2024). It's also reshaping construction by allowing large-scale component and structural fabrication, boosting efficiency and minimizing waste. In art and design, it empowers the creation of intricate pieces like sculptures and jewelry that are otherwise nearly impossible by hand (Choonara et al., 2016; Nadagouda et al., 2020).

8.2 3D Printing Production Techniques

3D printing technologies are generally categorized by material state into three main types: Powder-Based, Solid-Based, and Liquid-Based methods (Srinivasan et al., 2021).

8.2.1 Powder-Based Systems

Powder-based 3D printing involves using material powders and inkjet-style deposition. Key techniques include Laser Metal Deposition (LMD), Electron Beam Melting (EBM), Selective Laser Melting (SLM), and Direct Metal Laser Sintering (DMLS).

8.2.1.1 Laser Metal Deposition (LMD)

In Laser Metal Deposition (LMD), metal powder is injected into a laser-induced melt pool on a substrate, solidifying layer by layer to form 3D structures. It's widely used to restore parts like molds, valves, and screws. Composite materials can also be produced by blending metal or ceramic powders (Rumman et al., 2019).

8.2.1.2 Direct Metal Laser Sintering (DMLS)

Direct Metal Laser Sintering (DMLS) uses metal powder and laser energy guided by a CAD model to selectively melt layers on a build platform. Repeating this process forms intricate metal parts, with features as fine as 0.1 mm—ideal for complex, high-precision components (Maurya et al., 2021).

8.2.1.3 Selective Laser Melting (SLM)

Selective Laser Melting (SLM), similar to DMLS, fully melts metal powders using a high-powered laser to achieve near-fully dense parts (up to 99.9%). It's ideal for complex, high-performance components in aerospace and medical applications (Yap et al., 2015).

8.2.1.4 Selective Laser Sintering (SLS)

Selective Laser Sintering (SLS) is a mature industrial 3D printing method that builds parts layer by layer from a powder bed using a high-energy laser to selectively fuse particles. Unlike powder bed methods that use binders, SLS eliminates solvents, enabling faster production. A roller spreads fresh powder after each layer, and the process repeats until the object is complete. Once finished, the part is retrieved from the powder bed. Selective Laser Sintering (SLS) is a single-step process that doesn't require filament extrusion like FDM. Its laser precision enables higher-resolution prints, making it ideal for intricate designs. However, SLS relies on high-energy input to sinter metals, ceramics, or powdered plastics, which limits its use in pharmaceuticals due to potential thermal degradation of active ingredients (Duflou, 2019). Its ability to print complex geometries without support structures adds to its value. Similarly, Digital Light Processing (DLP) offers high-resolution, fast printing using a digital projector to cure resin—suited for detailed components.

8.2.1.5 Electron Beam Melting (EBM)

EBM uses a vacuum and high temperatures (up to 700°C) to melt metal powders with an electron beam, minimizing contamination and stress. It's widely applied for fabricating dense, high-strength metal parts (Sing et al., 2016).

8.2.2 Solid-Based Systems

Solid-based 3D printing uses materials like filaments, sheets, or pellets. Key methods include Wire Pulse Arc Additive Manufacturing (WPAAM), Electron Beam Freeform Fabrication (EBFF), and Fused Deposition Modeling (FDM).

8.2.2.1 Fused Deposition Modeling (FDM)

Fused Deposition Modeling (FDM), also known as Fused Filament Fabrication (FFF), has been a leading additive manufacturing technique for over two decades. It builds objects layer by layer from thermoplastic filaments using a computer-controlled system based on a digital 3D model. Key performance factors include material choice, print settings (e.g., raster angle, infill, print speed, extrusion temperature), and structural design, all of which significantly affect mechanical properties (Salmi, 2020; Cano-Vicent, 2021; Singh et al., 2020). FDM enables fabrication of solid and porous parts with customizable shapes and interconnectivity—ideal for rapid prototyping. Ongoing research focuses on improving resolution, addressing anisotropy, and enhancing surface finish.

8.2.2.2 Electron Beam Freeform Fabrication (EBFF)

Electron Beam Freeform Fabrication (EBFF) uses an electron beam to melt wire feedstock in a vacuum, ensuring clean processing and high material efficiency. It's ideal for producing large, complex metal parts with minimal material and energy use (Taminger and Hafley, 2003).

8.2.2.3 Wire Pulse Arc Additive Manufacturing (WPAAM)

Wire Pulse Arc Additive Manufacturing (WPAAM) uses an electric arc to melt metal wire, building parts layer by layer with high deposition efficiency and minimal waste. It's well-suited for producing large, complex metal components with high material use and cost efficiency (Sathish et al., 2021).

Powder-based techniques like SLM and EBM use high-energy lasers or electron beams to fully melt metal powders, producing near-net-shape parts with up to 99.9% density and excellent mechanical properties, including high tensile strength and fatigue resistance (Bi et al., 2021; Murr et al., 2020). In contrast, solid-based methods such as FDM and SLA use thermoplastic filaments or photocurable resins to build parts, offering different performance characteristics due to material and process differences. Comparative details of these systems are presented in Table 8.1.

Table 8.1: Comparison table for the Powder-Based and Solid-Based systems.

Aspect	**SLM/EBM (Powder-Based)**	**FDM/SLA (Solid-Based)**	**References**
Materials	Metals, alloys (Ti, Al, Inconel, HEAs)	Thermoplastics (FDM); Photopolymers (SLA)	(Li and Wang, 2024)
Mechanical Properties	High strength (900–1200 MPa), fatigue-resistant	Moderate (20–100 MPa), anisotropic (FDM), brittle (SLA)	(Li and Wang, 2024; Patel et al., 2024)
Thermal/ Electrical	High conductivity, heat-resistant	Low conductivity unless filled (FDM); Poor (SLA)	(Li and Wang, 2024; Patel, 2024; Zhang et al., 2023)
Resolution	Moderate (±50–100 μm), rougher surfaces (Ra 5–20 μm)	Low (FDM: ±200–500 μm); High (SLA: ±25–50 μm)	(Bi et al., 2021; Li and Wang, 2024; Patel et al., 2024)
Applications	Aerospace, biomedical, energy	Prototyping, medical, microfluidics, art	(Patel et al., 2024; Zhang et al., 2023; Huang et al., 2023; Li et al., 2024)
Cost	High ($100–$500/kg, $500k–$1M equipment)	Low (FDM: $20–$100/kg); Moderate (SLA: $50–$200/L)	(Peels, 2024; Li and Wang, 2024; Patel, 2024)
Speed	Slow (10–80 cm^3/h)	Moderate to fast (20–150 cm^3/h)	(Peels, 2024; Li and Wang, 2024; Patel, 2024)

Selective Laser Melting (SLM) and Electron Beam Melting (EBM) produce dense, high-performance metal components with excellent mechanical, thermal, and electrical properties, making them ideal for aerospace, biomedical, and energy applications—though high cost and post-processing remain challenges (Peels, 2024; Li et al., 2024). In contrast, Fused Deposition Modeling (FDM) and Stereolithography (SLA) are more affordable and faster, suited for prototyping and specialized uses like medical models, where resolution (SLA) or flexibility (FDM) matters. Although limited to polymers, advances in composite filaments and resins are expanding their functional range (Ma and Zhang, 2024; Huang et al., 2024). The choice between techniques hinges on performance goals, materials, and cost constraints—SLM and EBM for structural parts, FDM and SLA for precision or short-run needs.

8.2.3 Liquid-based Systems

Liquid-based 3D printing uses photosensitive resins that cure under light—typically from lasers or digital projectors. The main techniques are Stereolithography (SLA) and Digital Light Processing (DLP).

8.2.3.1 Stereolithography (SLA)

Stereolithography (SLA) uses a laser to cure liquid photopolymer resin layer by layer, forming 3D objects with high accuracy. It's preferred for producing finely detailed parts with smooth finishes due to its precision (DeSantis, 2019).

8.2.3.2 Digital Light Processing (DLP)

Digital Light Processing (DLP) speeds up printing by curing full resin layers simultaneously using a digital projector. With a digital micromirror device (DMD), it achieves high resolution and handles intricate geometries, making it efficient and versatile for detailed 3D printing tasks (Zhang et al., 2019).

3D printing technologies offer distinct benefits and drawbacks. Some emphasize affordability and material range, while others excel in precision and surface quality. Table 8.2 highlights key advantages and limitations of common techniques. Still, selecting the right method requires weighing trade-offs such as mechanical strength, material compatibility, and post-processing complexity (Rezaie et al., 2023).

8.3 Composite Materials and 3D Printing

Composite materials are engineered by combining two or more distinct phases—typically a matrix and a reinforcement—to achieve superior properties over individual components. The matrix (e.g., polymers, metals, ceramics) binds the structure and transfers loads, while reinforcements (e.g., carbon, glass, aramid

Table 8.2: Advantages and limitations of commonly used 3D printing techniques.

Technique	**Advantages**	**Limitations**
FDM (Fused Deposition Modeling)	- Cost-effective and low fire risk due to use of non-explosive, non-flammable materials. - Compatible with a wide range of materials.	- Limited resolution and accuracy. - Requires additional post-processing to refine surface finish.
DLP (Digital Light Processing)	- Produces smooth surfaces. - Simple and user-friendly setup.	- Limited suitability for printing surgical guides due to precision constraints.
SLA (Stereolithography)	- Offers high resolution, accuracy, and precision. - Supports a broad variety of printable materials.	- Relatively expensive. - Requires complex post-processing. - Results in mechanically weaker structures.
SLS (Selective Laser Sintering)	- Cost-effective equipment. - Enables functional prototyping using a wide range of materials.	- Requires powder-form starting materials. - Not suitable for large-scale parts. - Demands high maintenance.
SLM (Selective Laser Melting)	- Lower-cost machinery components.	- High melting temperatures required. - Risk of thermal shock during processing.

fibers, or particles) provide strength and stiffness. As shown in Table 8.3, these materials offer high strength-to-weight ratios, corrosion resistance, and durability, making them essential in sectors like aerospace, automotive, and construction. Despite their advantages, challenges remain in terms of cost, recyclability, and manufacturability. Ongoing advances aim to optimize design and processing to expand their applications further (Haghi et al., 2013; Rajak et al., 2019; Nagavally, 2017).

Composites are widely applied in aerospace, automotive, construction, sports, and healthcare due to their lightweight, high strength, and durability. In aerospace and automotive sectors, they improve fuel efficiency and safety. Construction uses composites in bridges, pipelines, and reinforcements for enhanced strength. Their low weight makes them ideal for sporting goods like bicycles and rackets. In medicine, composites support customized implants and prosthetics due to their biocompatibility (Nagavally, 2017; Bhong et al., 2023).

3D printing overcomes the limitations of conventional methods by enabling the fabrication of composite structures with complex, customized geometries. This precise control over material placement enhances strength, durability, and functionality—ideal for applications in aerospace, defense, automotive, and construction. Unlike traditional processes, 3D printing allows rapid, low-volume

Table 8.3: Comparative analysis of composite materials.

Composite type	Matrix material	Reinforcement	Tensile strength (MPa)	Density (g/cm³)	Thermal resistance	Typical applications	References
CFRP (Carbon Fiber Reinforced Polymer)	Epoxy Resin	Carbon Fiber	600–1500	1.6	High	Aerospace, Sports, Automotive	(Mallick, 2007; Hull and Clyne, 1996)
MMC (Metal Matrix Composite)	Aluminum	SiC/TiC Particulates	250–500	2.5–3.0	High	Aerospace, Engines, Brakes	(Clyne and Withers et al., 1995; Miracle, 2005)
GFRP (Glass Fiber Reinforced Polymer)	Polyester/ Epoxy	Glass Fiber	300–900	1.8–2.0	Moderate	Construction, Marine, Auto Parts	(Mallick, 2007; Strong, 2008)
Kevlar Composite	Epoxy	Aramid (Kevlar) Fiber	3000–3600	1.44	High	Defense, Helmets, Ropes	(Mouritz, 2007; Zhang et al., 2014)
Natural Fiber Composite	Polypropylene	Jute/Hemp/Flax	100–300	1.2–1.5	Low-Moderate	Automotive Interiors, Packaging	(Pickering et al., 2016; Mohanty et al., 2000)
CMC (Ceramic Matrix Composite)	Silicon Carbide	Carbon/SiC Fibers	300–500	2.5–3.2	Very High	Turbines, Spacecraft, Furnaces	(Naslain, 2004; Chawla, 2012)

production of tailored parts with minimal waste and unmatched design flexibility, making it well-suited for solving intricate engineering challenges efficiently. 3D printing overcomes the limitations of conventional methods by enabling the fabrication of composite structures with complex, customized geometries. This precise control over material placement enhances strength, durability, and functionality—ideal for applications in aerospace, defense, automotive, and construction. Unlike traditional processes, 3D printing allows rapid, low-volume production of tailored parts with minimal waste and unmatched design flexibility, making it well-suited for solving intricate engineering challenges efficiently (Smith and Lee, 2023).

Sustainability is a major advantage of 3D printing in composite production. As an additive process, it reduces material waste by depositing material only where needed. The use of bio-based and recyclable materials further lessens reliance on non-renewable resources, supporting closed-loop, eco-friendly manufacturing. 3D printing accelerates product development by enabling quick prototyping and easy design modifications, significantly shortening the concept-to-product cycle. On-demand production also reduces lead times and inventory needs, boosting flexibility and responsiveness to shifting demands.

Another key advantage of 3D printing is precise control over material placement, enabling optimization of structural features like strength-to-weight ratios and other mechanical properties. This control supports the fabrication of functionally graded composites with spatially tailored stiffness and thermal conductivity, enhancing performance in targeted applications (Tabassum et al., 2023).

8.3.1 Application of 3D Printing Composite Material

3D printed composite biomaterials offer key advantages for biomedical applications, particularly in tissue engineering. They combine mechanical strength, biocompatibility, and the ability to incorporate tissue-like structures. Scaffolds—typically porous and interconnected—are 3D printed to support nutrient diffusion, vascularization, and mechanical stability. Inkjet-based printing is often used to fabricate calcium phosphate scaffolds, bonding powder with a polymer and sintering for strength. Tetracalcium phosphate combined with biodegradable fillers like β-TCP enables complex geometries. However, challenges include difficulty in forming sub-1 mm channels, improving compressive strength, and reducing microporosity. To address this, TCP is blended with HA, calcium sulfate, or PCL to enhance scaffold performance.

Fielding et al. (2012) improved the mechanical and biological performance of calcium phosphate scaffolds by adding ZnO and SiO_2 to the powder. These composites, along with pure tricalcium phosphate scaffolds, were produced via

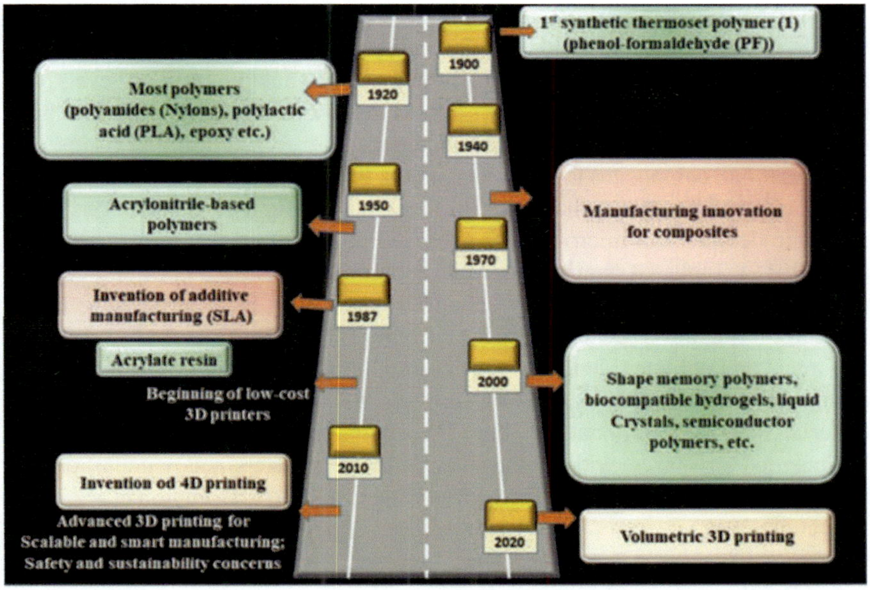

Figure 8.1: Confinement and development of 3D Printing.

inkjet printing and sintered at 1250°C to ensure durable bonding (Kalsoom et al., 2016; Khalyfa et al., 2007). Figure 8.1 illustrates the century-long evolution of 3D printing.

3D printing, particularly inkjet-based methods, addresses the challenges of incorporating carbon nanotubes (CNTs) into composites by enabling precise deposition using conductive inks made from well-dispersed CNTs via biopolymer or chemical modification. This has led to CNT-based films with good electrical conductivity. Additionally, composites incorporating carbon black or silver nanoparticles support the creation of flexible circuits, strain sensors, and optoelectronic devices—serving as low-cost alternatives to indium tin oxide (ITO). Emerging developments include piezoelectric nanoparticle composites and graphene-based scaffolds for next-gen printed electronics (Zhi et al., 2015).

Inkjet printing is gaining traction in electrochromic device fabrication due to its ability to produce high-resolution patterns with minimal material waste. However, performance is still hindered by ink viscosity and surface tension affecting droplet formation. Advances using low-viscosity colloidal composites of intrinsically conductive polymers (ICPs)—like quantum dot-polymer blends—enable luminescent, energy-efficient films. Additionally, ICP inks such as polyaniline–silica and PEDOT–silica composites have facilitated the development of color-changing electrochromic displays in multicolored AC systems (Shim et al., 2008; Tekin et al., 2007; Hwang et al., 2015).

3D printing has advanced the development of thermally functional composites tailored for heat management. For example, adding graphene oxide (GO) to ABS via FDM improves thermal conductivity, though high filler levels can cause nozzle clogging. SLA printing avoids this by curing resin with UV light, enabling use of thermally enhanced materials like TiO_2-reinforced resins and microdiamond composites for heat dissipation. Photo-curable resins incorporating thermoelectric materials like $Bi_{0.5}Sb_{1.5}Te_3$ further support applications such as bionic devices and micro-batteries (Huang et al., 2015; Wei et al., 2015; He et al., 2015). Table 8.4 summarizes key properties and uses. 3D printing enables the development of mechanically enhanced composites by incorporating fillers like carbon fibers and SiC whiskers into polymers. Techniques such as *in situ* fiber alignment and polymer blending improve toughness and strength. The integration of shape memory alloys (SMAs) supports the creation of smart, adaptive materials. Unlike traditional methods, 3D printing allows fabrication of complex bio-inspired structures, with tuning via printing parameters and nanoparticles to further boost flexibility and performance (Ning et al., 2015; Compton et al., 2014; Ahn et al., 2012).

8.4 Challenges and Future Perspectives

3D printing faces key challenges including material limitations, scalability, high costs, and post-processing needs (Singh & Gupta, 2024; Li & Wang, 2024). Many printable materials lack the strength or durability of conventional ones. Producing large parts is time-intensive and constrained by printer size. High-end printers and specialized materials also raise costs, though these are gradually declining.. Post-processing—such as sanding or curing—is often required for functional or aesthetic quality, adding time and labor. Meanwhile, emerging areas like bioprinting and multi-material printing are expanding capabilities, enabling tissue fabrication and multifunctional components (Huang et al., 2024; Ma & Zhang, 2024). Sustainability is gaining momentum, with a focus on recyclable and bio-derived materials to lower environmental impact (Saqib et al., 2024; Shim et al., 2008).

8.5 Conclusion

Composite materials in 3D printing are unlocking new possibilities across industries by enhancing the functional scope of additive manufacturing. Material selection depends on application-specific needs like strength, weight, heat resistance, and cost. While sustainable natural fiber composites are rising, high-performance sectors still rely on CFRPs and CMCs for their superior properties. Despite challenges such as cost, processing complexity, and equipment limits,

Table 8.4: An overview of the main characteristics and uses of 3D printable composite materials.

Approach	Properties of 3D-printed composites	Applications	References
Applications in Biomedical	Young's modulus of 80 GPa, bending strength of 80 MPa, and decreased and isotropic shrinkage in the composite	Tissue and bone replacement	(Winkel et al., 2012)
	53.1% porosity, 93.9 g cm^{-3} bulk density, and improved biocompatibility	Substitutes of bone	(Chi et al., 2014)
	Biocompatible, high mechanical strength, degradable	Implants of bone	(Inzana et al., 2014)
	55% porosity, 0.12–0.88 MPa compressive strength, and enhanced biological characteristics such as osteoconduction and osteointegration	Bone tissue engineering	(Cox et al., 2015)
Mechanical properties enhancement	Improved mechanical properties	Completely operational end-use components	(Goodridge et al., 2011)
	Rapid prototyping of near net forms and dimensions, ultra-high melting metal composites (>2500 ºC), and porosity of 57.9%	Refractory rocket nozzle, spacecraft	(Lipke et al., 2010)
	Shear stress of 24 MPa, strain of 2%	High-quality mechanical parts and materials	(Christiyan et al., 2016)
Bending/ twisting strength enhancement	reduced porosity (around 20%) and Young's modulus compared to the compressed samples	Biocomposites	(Le Duigou et al., 2016)
	Significant deformations and intricately molded shapes	Morphing for overwhelming the geometric constraints of constructions or actuators	(Ahn et al., 2009)
Composite materials that are electrically active	With an optical transparency of 85% and a sheet resistance of 100 $k\Omega^{-1}$, it is electrically conductive	Electronics	(Small, 2007)
	Resistivity is ¼ 0.0960.01 ohm m^{-1} in-plane with the layers and ¼ 0.1260.01 ohm m^{-1} perpendicular to the layers.	Capacitive buttons, electronic sensors	(Leigh et al., 2012)

Table 8.4 contd. ...

...Table 8.4 contd.

Approach	Properties of 3D-printed composites	Applications	References
Dielectric material applications	Electrical conductivity of 18 Ω sq^{-1}	Paper electrode that is lightweight and flexible	(Chi et al., 2014)
	Enhanced permittivity and a low tangent value for dielectric loss	Insulating components and spacers	(Kurimoto et al., 2015)
Thermal properties enhancement	Conductivity of heat of 0.2 W m^{-1} K^{-1}	Thermoelectric material	(He et al., 2015)
	27.4% porosity and enhanced thermal stability	Rocket motors	(Huang et al., 2015)
Optical properties enhancement	Luminous effectiveness improved optical characteristics for the blue phosphor powered by 420 V at 31 kHz, 0.1 lm W^{-1}	Luminescent light-transforming thin films for AC	(Wood et al., 2009)
	Optically active	Electrochromic devices	(Tekin et al., 2007)

advances in bio-printing, multi-material systems, and sustainable methods are driving future growth. Ongoing innovations in materials and printing technology promise broader adoption and more sophisticated applications.

References

Ahn, S., Hong, J.Y., Hong, M.K., Jang, Y.P., Oh, M.S. et al. 2009. Structural determination of sildenafil and its analogues in dietary supplements by fast-atom bombardment collision-induced dissociation tandem mass spectrometry. *Rapid Communications in Mass Spectrometry: An International Journal Devoted to the Rapid Dissemination of Up-to-the-Minute Research in Mass Spectrometry*, 23(19): 3158–3166. https://doi.org/10.1002/rcm.4238.

Ahn, S.H., Lee, K.T., Kim, H.J., Wu, R., Kim, J.S. et al. 2012. Smart soft composite: An integrated 3D soft morphing structure using bend-twist coupling of anisotropic materials. *International Journal of Precision Engineering and Manufacturing*, 13(4): 631–634. https://doi.org/10.1007/s12541-012-0081-8.

Barnatt, C. 2013. 3D printing: The next industrial revolution. *Explaining the Future*, 276. http://explainingthefuture.com/3dp2e.html.

Bhong, M., Khan, T.K., Devade, K., Krishna, B.V., Sura, S. et al. 2023. Review of composite materials and applications. *Materials Today: Proceedings*. https://doi.org/10.1016/j.matpr.2023.10.026.

Bi, J., Liu, Y. and Zhang, X. 2021. Microstructure and mechanical properties of SLM Ti-6Al-4V. *Additive Manufacturing*, 46: 102–115. doi:10.1016/j.addma.2021.102115.

Cano-Vicent, A., Tambuwala, M.M., Hassan, S.S., Barh, D., Aljabali, A.A. et al. 2021. Fused deposition modelling: Current status, methodology, applications and future prospects. *Additive Manufacturing*, 47: 102378. https://doi.org/10.1016/j.addma.2021.102378.

Chawla, K.K. 2012. *Composite Materials: Science and Engineering*. Springer.

Chi, K., Zhang, Z., Xi, J., Huang, Y., Xiao, F. et al. 2014. Freestanding graphene paper supported three-dimensional porous graphene–polyaniline nanocomposite synthesized by inkjet printing and in flexible all-solid-state supercapacitor. *ACS Applied Materials & Interfaces*, 6(18): 16312–16319. https://doi.org/10.1021/am504539k.

Choonara, Y.E., du Toit, L.C., Kumar, P., Kondiah, P.P., Pillay, V. et al. 2016. 3D-printing and the effect on medical costs: A new era? *Expert Review of Pharmacoeconomics and Outcomes Research*, 16(1): 23–32. https://doi.org/10.1586/14737167.2016.1138860 .

Christiyan, K.J., Chandrasekhar, U. and Venkateswarlu, K. 2016. A study on the influence of process parameters on the mechanical properties of 3D printed ABS composite. In *IOP Conference Series: Materials Science and Engineering*, 114: 012109. https://doi.org/10.1088/1757-899x/114/1/012109.

Clyne, T.W. and Withers, P.J. 1995. *An Introduction to Metal Matrix Composites*. Cambridge University Press.

Compton, B.G. and Lewis, J.A. 2014. 3D-printing of lightweight cellular composites. *Advanced Materials*, 26(34): 5930–5935. https://doi.org/10.1002/adma.201401804.

Cox, S.C., Thornby, J.A., Gibbons, G.J., Williams, M.A., Mallick, K.K. et al. 2015. 3D printing of porous hydroxyapatite scaffolds intended for use in bone tissue engineering applications. *Materials Science and Engineering: C*, 47: 237–247. https://doi.org/10.1016/j.msec.2014.11.024.

DeSantis, J.D. 2019. Stereolithography: The future of additive manufacturing. *Journal of Manufacturing Processes*.

Duflou, C.D.R. 2019. Selective laser sintering of metals and polymers. *Rapid Prototyping Journal*.

Goodridge, R.D., Shofner, M.L., Hague, R.J., McClelland, M., Schlea, M.R. et al. 2011. Processing of a polyamide-12/carbon nanofibre composite by laser sintering. *Polymer Testing*, 30(1): 94–100. https://doi.org/10.1016/j.polymertesting.2010.10.011.

Haghi, A.K., Oluwatobi Samuel Oluwafemi, Josmin P. Jose and Hanna J. Maria (eds.). 2013. *Composites and Nanocomposites*. CRC Press, 199. https://doi.org/10.1201/b14994.

He, M., Zhao, Y., Wang, B., Xi, Q., Zhou, J. et al. 2015. 3D printing fabrication of amorphous thermoelectric materials with ultralow thermal conductivity. *Small*, 11(44): 5889–5894. https://doi.org/10.1002/smll.201502153.

Huang, C., Jian, G., DeLisio, J.B., Wang, H., Zachariah, M.R. et al. 2015. Electrospray deposition of energetic polymer nanocomposites with high mass particle loadings: A prelude to 3D printing of rocket motors. *Advanced Engineering Materials*, 17(1): 95–101. https://doi.org/10.1002/adem.201400151.

Huang, S., Wei, H. and Li, D. 2024. Advancements in 3D printing techniques for biomedical applications: A comprehensive review. *Discover Materials*. doi:10.1007/s43939-024-00073-9.

Hull, D. and Clyne, T.W. 1996. *An Introduction to Composite Materials* (2nd ed.). Cambridge University Press.

Hwang, S., Reyes, E.I., Moon, K.S., Rumpf, R.C., Kim, N.S. et al. 2015. Thermo-mechanical characterization of metal/polymer composite filaments and printing parameter study for fused deposition modeling in the 3D printing process. *Journal of Electronic Materials*, 44: 771–777. https:// doi:10.1007/s11664-014-3425-6.

Inzana, J.A., Olvera, D., Fuller, S.M., Kelly, J.P., Graeve, O.A. et al. 2014. 3D printing of composite calcium phosphate and collagen scaffolds for bone regeneration. *Biomaterials*, 35(13): 4026–4034. https://doi.org/10.1016/j.biomaterials.2014.01.064.

Joshi, S.C. and Sheikh, A.A. 2015. 3D printing in aerospace and its long-term sustainability. *Virtual and Physical Prototyping*, 10(4): 175–185. https://doi.org/10.1080/17452759.2015.1111519.

Kalsoom, U., Nesterenko, P.N. and Paull, B. 2016. Recent developments in 3D printable composite materials. *RSC Advances*, 6(65): 60355–60371. https://doi.org/10.1039/C6RA11334F.

Khalyfa, A., Vogt, S., Weisser, J., Grimm, G., Rechtenbach, A. et al. 2007. Development of a new calcium phosphate powder-binder system for the 3D printing of patient specific implants. *Journal of Materials Science: Materials in Medicine*, 18(5): 909–916. https://doi.org/10.1007/s10856-006-0073-2.

Kulkarni, A. and Sharma, R. 2024. Revolutionising fabrication: Advances and applications of 3D printing with composite materials. *Journal of Composite Materials*. doi:10.1080/00219983.2024.2336160M.

Kurimoto, M., Yamashita, Y., Ozaki, H., Kato, T., Funabashi, T. et al. 2015. 3D printing of conical insulating spacer using alumina/UV-cured-resin composite. In *2015 IEEE Conference on Electrical Insulation and Dielectric Phenomena (CEIDP)*, 463–466. https://doi.org/10.1109/CEIDP.2015.7352047.

Le Duigou, A., Castro, M., Bevan, R. and Martin, N. 2016. 3D printing of wood fibre biocomposites: From mechanical to actuation functionality. *Materials & Design*, 96: 106–114. https://doi.org/10.1016/j.matdes.2016.02.018.

Leigh, S.J., Bradley, R.J., Purssell, C.P., Billson, D.R., Hutchins, D.A. et al. 2012. A simple, low-cost conductive composite material for 3D printing of electronic sensors. *PloS One*, 7(11): e49365. https:// doi:10.1371/journal.pone.0049365 .

Li, X. and Wang, Z. 2024. Non-covalent functionalization for MWCNT dispersion in stereolithography. *Additive Manufacturing*, 82: 103–115. doi:10.1016/j.addma.2023.12.008.

Li, Y., Zhang, Q. and Chen, L. 2024. Lattice-structured Inconel 718 via SLM for aerospace applications. *Journal of Materials Processing Technology*, 325: 118–130. doi:10.1016/j.jmatprotec.2023.118.

Lipke, D.W., Zhang, Y., Liu, Y., Church, B.C., Sandhage, K.H. et al. 2010. Near net-shape/net-dimension ZrC/W-based composites with complex geometries via rapid prototyping and displacive compensation of porosity. *Journal of the European Ceramic Society*, 30(11): 2265–2277. https://doi.org/10.1016/j.jeurceramsoc.2010.01.011.

Ma, Q. and Zhang, Y. 2024. Advances in 3D printing for polymer composites: A review. *InfoMat*, 6(3): e12345. doi:10.1002/inf2.12345.

Mallick, P.K. 2007. *Fiber-reinforced Composites: Materials, Manufacturing, and Design* (3rd ed.). CRC Press.

Maurya, N.K., Sharma, R., Kumar, N., Kumar, A., Anand, P. et al. 2021. An overview of investigation of Fatigue, tensile strength and hardness of the components fabricated through direct metal laser sintering (DMLS) process. *Materials Today: Proceedings*, 47: 3979–3984. https://doi.org/10.1016/j.matpr.2021.04.131.

Miracle, D.B. 2005. Metal matrix composites—From science to technological significance. *Composites Science and Technology*, 65(15-16): 2526–2540.

Mohanty, A.K., Misra, M. and Drzal, L.T. 2000. Sustainable bio-composites from renewable resources: Opportunities and challenges in the green materials world. *Journal of Polymers and the Environment*, 10(1): 19–26.

Mouritz, A.P. 2007. *Introduction to Aerospace Materials*. Woodhead Publishing.

Murr, L.E., Gaytan, S.M. and Martinez, E. 2020. Electron beam melting of advanced materials and structures. *Materials Science and Engineering: A*, 773: 138–150. doi:10.1016/j.msea.2019.138150.

Nadagouda, M.N., Rastogi, V. and Ginn, M. 2020. A review on 3D printing techniques for medical applications. *Current Opinion in Chemical Engineering*, 28: 152–157. https://doi.org/10.1016/j.coche.2020.05.007.

Nagavally, R.R. 2017. Composite materials-history, types, fabrication techniques, advantages, and applications. *Int. J. Mech. Prod. Eng.*, 5(9): 82–87. https://www.digitalxplore.org/up_proc/pdf/240-146959633425-30.pdf.

Naslain, R. 2004. Design, preparation and properties of non-oxide CMCs for application in engines and nuclear reactors: An overview. *Composites Science and Technology*, 64(2): 155–170.

Ning, F., Cong, W., Qiu, J., Wei, J., Wang, S. et al. 2015. Additive manufacturing of carbon fiber reinforced thermoplastic composites using fused deposition modeling. *Composites Part B: Engineering*, 80: 369–378. https://doi.org/10.1016/j.compositesb.2015.06.013.

Patel, V., Gupta, S. and Sharma, M. 2024. High-conductivity PLA/MWCNT nanocomposites for 3D printing. *Materials & Design*, 238: 112567. doi:10.1016/j.matdes.2023.112567.

Peels, J. 2024. 3D printing predictions for 2025: Metal 3D printing. 3DPrint.com. Published December 27.

Pickering, K.L., Efendy, M.G.A. and Le, T.M. 2016. A review of recent developments in natural fibre composites and their mechanical performance. *Composites Part A: Applied Science and Manufacturing*, 83: 98–112.

Rajak, D.K., Pagar, D.D., Menezes, P.L. and Linul, E. 2019. Fiber-reinforced polymer composites: Manufacturing, properties, and applications. *Polymers*, 11(10): 1667. https://www.mdpi.com/2073-4360/11/10/1667.

Rezaie, F., Farshbaf, M., Dahri, M., Masjedi, M., Maleki, R. et al. 2023. 3D printing of dental prostheses: Current and emerging applications. *Journal of Composites Science*, 80. https://doi.org/10.3390/jcs7020080.

Rumman, R., Lewis, D.A., Hascoet, J.Y. and Quinton, J.S. 2019. Laser metal deposition and wire arc additive manufacturing of materials: An overview. *Archives of Metallurgy and Materials*, 467–473. https://doi.org/10.24425/amm.2019.127561.

Salmi, A.J. 2020. Fused deposition modeling: A review. *Advances in Manufacturing Science and Technology*.

Saqib, M.A., Abbas, M.S., Tanaka, H. et al. 2024. Sustainability and innovation in 3D printing: Outlook and trends. *Clean Technologies and Recycling*, 4(1): 1–21. doi:10.3934/ctr.2024001.

Sathish, T., Tharmalingam, S., Mohanavel, V., Ashraff Ali, K.S., Karthick, A. et al. 2021. Weldability investigation and optimization of process variables for TIG-welded aluminium alloy (AA 8006). *Advances in Materials Science and Engineering*, 1(2021): 2816338. https://doi.org/10.1155/2021/2816338.

Shahrubudin, N., Lee, T.C. and Ramlan, R.J. 2019. An overview on 3D printing technology: Technological, materials, and applications. *Procedia Manufacturing*, 35: 1286–1296. https://doi.org/10.1016/j.promfg.2019.06.089.

Shim, G.H., Han, M.G., Sharp-Norton, J.C., Creager, S.E., Foulger, S.H. et al. 2008. Inkjet-printed electrochromic devices utilizing polyaniline–silica and poly (3, 4-ethylenedioxythiophene)–silica colloidal composite particles. *Journal of Materials Chemistry*, 18(5): 594–601. https://doi.org/10.1039/B712766A.

Sing, S.L., An, J., Yeong, W.Y. and Wiria, F.E. 2016. Laser and electron-beam powder-bed additive manufacturing of metallic implants: A review on processes, materials and designs. *Journal of Orthopaedic Research*, 34(3): 369–385. https://doi.org/10.1002/jor.23075.

Singh, S. and Gupta, D. 2024. Large-format additive manufacturing of polymers: A review of fabrication processes, materials, and design. *Journal of Manufacturing Technology*. doi:10.1080/10426914.2024.2336160.

Singh, S., Singh, G., Prakash, C. and Ramakrishna, S. 2020. Current status and future directions of fused filament fabrication. *Journal of Manufacturing Processes*, 55: 288–306. https://doi.org/10.1016/j.jmapro.2020.04.049.

Small, W.R. in het Panhuis M. 2007. Inkjet printing of transparent, electrically conducting single-walled carbon-nanotube composites. Small, 3(9): 1500–1503. https://doi.org/10.1002/smll.200700110.

Smith, J.A. and Lee, K. 2023. Advances in additive manufacturing: Material innovation through 3D printing. *Journal of Materials Science & Technology*, 45(3): 123–135. https://doi.org/10.1016/j.jmst.2023.01.012).

Srinivasan, D., Meignanamoorthy, M., Ravichandran, M., Mohanavel, V., Alagarsamy, S.V. et al. 2021. 3D printing manufacturing techniques, materials, and applications: An overview. *Advances in Materials Science and Engineering*, 1(2021): 5756563. https://doi.org/10.1155/2021/5756563.

Strong, A.B. 2008. *Fundamentals of Composites Manufacturing: Materials, Methods and Applications* (2nd ed.). SME.

Tabassum, T. and Mir, A.A. 2023. A review of 3d printing technology—The future of sustainable construction. *Materials Today: Proceedings*, 93: 408–414. https://doi.org/10.1016/j.matpr.2023.08.013.

Taminger, K. and Hafley, R.A. 2003. Electron beam freeform fabrication: A rapid metal deposition process. In *3rd Annual Automotive Composites Conference*. Troy, MI, USA, 1–6. https://ntrs.nasa.gov/citations/20040042496.

Tekin, E., Smith, P.J., Hoeppener, S., Van Den Berg, A.M., Susha, A.S. et al. 2007. Inkjet printing of luminescent CdTe nanocrystal–polymer composites. *Advanced Functional Materials*, 17(1): 23–28. https://doi.org/10.1002/adfm.200600587.

Wei, X., Li, D., Jiang, W., Gu, Z., Wang, X., Zhang, Z. and Sun, Z. 2015. 3D printable graphene composite. *Scientific Reports*, 5(1): 1–7. https:// doi:10.1038/srep11181.

Winkel, A., Meszaros, R., Reinsch, S., Muller, R., Travitzky, N. et al. 2012. Sintering of 3D-printed Glass/HAp composites. *J. Am. Ceram. Soc.*, 95: 3387–3393. https://doi.org/10.1111/j.1551-2916.2012.05368.x.

Wood, V., Panzer, M.J., Chen, J., Bradley, M.S., Halpert, J.E. et al. 2009. Quantum dot–polymer composites for displays: Inkjet-printed quantum dot–polymer composites for full-color AC-driven displays. *Advanced Materials*, 21(21): 1–5. https://doi.org/10.1002/adma.200990078.

Yap, C.Y., Chua, C.K., Dong, Z.L., Liu, Z.H., Zhang, D.Q. et al. 2015. Review of selective laser melting: Materials and applications. *Applied Physics Reviews*, 2(4): 041101. https://doi.org/10.1063/1.4935926.

Zhang, H., Zhao, Y. and Li, Q. 2023. High-entropy alloys via selective laser melting. *Materials Today*, 62: 45–58. doi:10.1016/j.mattod.2023.01.012.

Zhang, J., Hu, Q., Wang, S., Tao, J., Gou, M. et al. 2019. Digital light processing based three-dimensional printing for medical applications. *International Journal of Bioprinting*, 6(1): 242. https://doi.org/10.18063/ijb.v6i1.242.

Zhang, J., Wang, Y. and Zhu, B. 2014. Properties and applications of Kevlar fiber reinforced composites. *Advanced Materials Research*, 852: 103–107.

Zhi, X., Zhang, H.B., Liao, Y.F., Hu, Q.H., Gui, C.X. et al. 2015. Electrically conductive polycarbonate/carbon nanotube composites toughened with micron-scale voids. *Carbon*, 82: 195–204. https://doi.org/10.1016/j.carbon.2014.10.062.

CHAPTER

9 The Rise of Automation

Streamlining Composite Material Manufacturing

Ravi Shankar Rai,[1,*] *Alok Kumar*[2]
and *Chandrmani Yadav*[3]

Over recent years, automation has transformed manufacturing by replacing manual, labor-intensive processes with mechanized systems. Composite material production, traditionally costly and inconsistent, is particularly well-suited for automation. This chapter explores how technologies like robotics and machine learning enhance key processes—material handling, layup, inspection, and curing—leading to better efficiency, quality, and cost-effectiveness. It also highlights the integration of IoT and Industry 4.0 solutions, enabling smarter workflows and predictive maintenance for optimized composite manufacturing.

9.1 Introduction

9.1.1 Overview of Automation in Modern Manufacturing

Automation is reshaping modern manufacturing, particularly in precision-driven sectors like composite material (CM) production. Technologies such as robotics, AI, ML, and control systems streamline industrial workflows, enhancing

[1] Department of Automation and Robotics Engineering, JSPM's Rajarshi Shahu College of Engineering, Pune, India.
[2] Department of Mechanical Engineering, Indian Institute of Technology, Jodhpur, India.
[3] Marwadi University Research Center, Faculty of Engineering & Technology, Department of Mechanical Engineering, Marwadi University, Rajkot, Gujarat, India.
* Corresponding author: rash.mait@gmail.com
https://orcid.org/0000-0001-7406-2229

DOI: 10.1201/9781003607359-9

consistency, efficiency, and accuracy while reducing waste and costs (Patel et al., 2024). In CM manufacturing, automation addresses key challenges of manual processes—like inconsistent quality, labor costs, and inefficiency—through high-precision tasks such as fiber placement, resin infusion, cutting, and curing. When paired with machine vision, robotics ensures accurate alignment and minimizes defects (Rojek et al., 2022; Malaga et al., 2022). AI and ML enable predictive maintenance and process optimization by analyzing data to anticipate failures and allocate resources effectively. IoT integration further supports real-time data exchange across systems, fostering smart, interconnected production environments. While challenges like high initial investment, workforce reskilling, and job displacement remain, the long-term benefits—higher productivity, product quality, and operational savings—make automation a core pillar of advanced manufacturing (Javaid et al., 2022).

9.1.2 Importance of CM in Advanced Industries

Composite materials (CM), formed by combining distinct constituents with unique properties, are increasingly vital in advanced industries for their strength, durability, and design versatility. Their standout advantage—lightweight construction—enhances efficiency in aerospace, automotive, and renewable energy sectors by improving fuel economy and reducing emissions (Zhang et al., 2022; Ryan et al., 2024). For instance, CM-based wind turbine blades and aircraft components boost energy output and structural performance while resisting corrosion and harsh environments. Composites also excel in defense and maritime applications for their impact resistance and saltwater durability. Their tunable design—via fiber type, orientation, or matrix—enables customized solutions in architecture, sports gear, and biomedical devices. Moreover, bio-based and recyclable CMs are gaining traction as industries pursue sustainable, low-maintenance alternatives (Hu et al., 2023; Alves de Campos et al., 2022). Overall, CM innovation is reshaping product design and manufacturing, positioning them as essential enablers of future-ready technologies.

9.1.3 Automation in Transforming CM Manufacturing

Automation has transformed composite material (CM) production by enhancing efficiency, precision, and scalability. Key processes like fiber placement, resin infusion, curing, and trimming now rely on advanced sensors, robotics, and ML algorithms for improved speed and consistency. Notably, robotic systems such as automated tape laying (ATL) and automated fiber placement (AFP) ensure precise tape deposition and uniform alignment, reducing human error and improving product repeatability (Jayasekara et al., 2022). Automation in composite manufacturing now integrates advanced systems beyond robotics, including sensors and machine vision to monitor alignment and curing in real time. This data-driven approach helps maintain quality, reduce waste, and detect issues early. AI and ML further optimize operations through predictive maintenance and adaptive process control, adjusting settings based on material behavior to maintain consistency (Choi et al., 2023). Digital twins (DT) and IoT technologies enhance connectivity between machines and streamline workflows, enabling virtual simulations before actual execution. As demand grows—particularly in aerospace, automotive, and renewable energy—automation supports scalable, high-precision production.While initial costs and skilled labor needs remain challenges, the long-term gains in efficiency, product quality, and competitiveness outweigh these hurdles (Frketic et al., 2017). Continued development promises further innovation, expanding automation's role in creating advanced, high-performance composites.

9.2 Overview of CM

9.2.1 CM: Definition and Types

Composites are engineered by combining two or more distinct materials—typically a reinforcement and a matrix—each retaining its properties but working together to enhance overall performance. This synergy yields materials with superior mechanical, thermal, and chemical properties. Their lightweight, strength, corrosion resistance, and design flexibility make them essential in aerospace, automotive, construction, maritime, sports, and renewable energy sectors (Hsissou et al., 2021). Figure 9.1 classifies composites by matrix type. Polymer matrix composites (PMCs) are widely used due to low cost and adaptability, typically combining polymers like epoxy or thermoplastics with

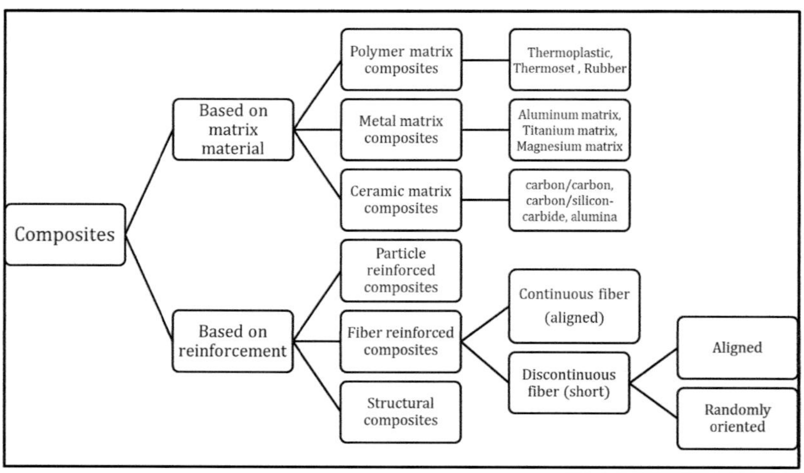

Figure 9.1: Categorization of CM and their components.

reinforcements such as carbon, glass, or aramid fibers—ideal for aerospace and automotive use. Metal matrix composites (MMCs) use metals (e.g., aluminum, titanium) reinforced with ceramic or metallic fibers for enhanced strength and wear resistance in high-performance applications (Bhong et al., 2023; Naik et al., 2023).

Metal matrix composites (MMCs), typically using matrices like magnesium, titanium, or aluminium with reinforcements such as silicon carbide or boron fibers, are well-suited for high-performance applications in automotive, aerospace, and defense. They offer outstanding strength, thermal conductivity, and wear resistance.

Ceramic matrix composites (CMCs), reinforced with carbon fibers, are known for their exceptional heat and weather resistance, making them ideal for high-temperature applications in automotive, aerospace, and gas turbines. Carbon–carbon composites, offering superior heat tolerance and strength-to-weight ratios, are widely used in thermal protection systems and rocket nozzles. Hybrid composites, which combine multiple reinforcement types within a single matrix, are designed to balance mechanical, thermal, and electrical properties for demanding applications (Talib and Hussam, 2022). Natural fiber composites—using jute, flax, hemp, or bamboo in polymer matrices—are valued for their lightweight and eco-friendly nature. They're increasingly used in construction, packaging, and vehicle interiors as demand for sustainable materials grows. High-performance composites remain key in aerospace and automotive sectors due to their superior strength and adaptability. Overall, composites offer tailored properties by combining diverse components, while automation has boosted their quality, efficiency, and expanded their industrial applications.

9.2.2 Properties and Advantages of CM

Composite materials (CM) offer an exceptional combination of strength, durability, and adaptability, making them ideal for industrial and advanced manufacturing. Their superior mechanical, thermal, and chemical properties stem from the synergy between the matrix and reinforcement components (Ali and Andriyana, 2020; Kumar et al., 2024). With a high strength-to-weight ratio, CM are widely used in aerospace and automotive sectors for lightweight yet robust components. Their resistance to corrosion and chemicals supports applications in marine and chemical industries, extending service life and reducing maintenance. CMs also provide design flexibility—adjusting fiber type, orientation, and volume enables tailored structural solutions. Their excellent thermal and electrical insulation makes them suitable for electronics and aviation heat management (Wood et al., 2022). Further, CM exhibit high fatigue and impact resistance, essential in vehicles and sports gear, along with minimal thermal expansion, ensuring dimensional stability in precision equipment (Peças et al., 2018). Altogether, these traits make CM superior to many conventional materials across performance-critical applications.

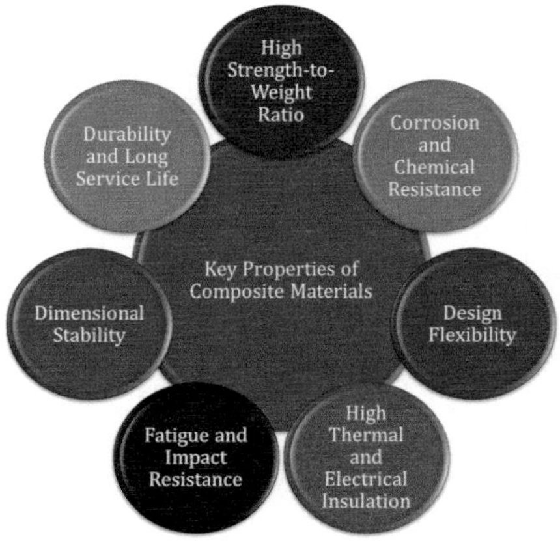

Figure 9.2: Key properties of CM.

Composite materials (CM) play a vital role in modern transportation due to their lightweight nature, improving fuel efficiency and lowering carbon emissions—key drivers of sustainability. Their high strength and rigidity support innovative structural designs with superior load-bearing capacity. Advancements in bio-based and recyclable composites further align with eco-friendly goals, while their durability minimizes maintenance and waste impacts. CM also enhance passenger safety by absorbing vibrations and impacts, making them ideal for vehicles, aircraft, and transit systems. Automation in CM manufacturing boosts production efficiency, cuts costs, and improves consistency, enabling tailored solutions in aerospace, defense, and medical sectors (Namsheer and Rout, 2021; Drissi et al., 2019). In essence, CM have become foundational to advanced manufacturing, and their usage is expected to expand as automation drives future innovation.

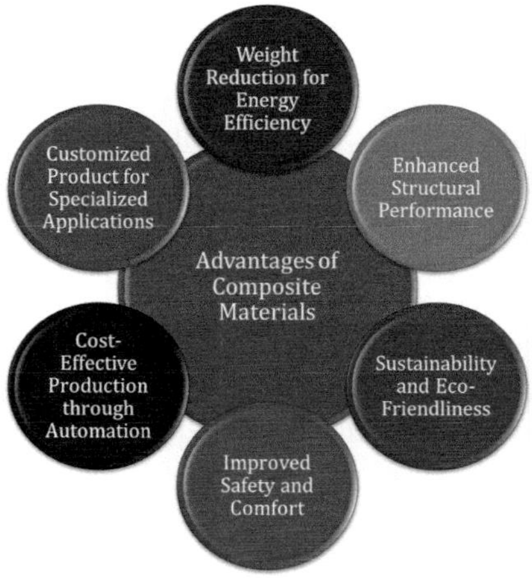

Figure 9.3: Advantages of CM.

9.2.3 Applications for CM in Industries

Composite materials (CM) offer high strength, low weight, corrosion resistance, and design flexibility—making them essential across industries prioritizing performance and innovation. Their use is rapidly expanding due to advances in automated manufacturing.

- Aerospace: CM are widely used in wings, fuselages, and tail sections for their strength-to-weight ratio and fatigue resistance, supporting fuel efficiency, increased cargo capacity, and reduced emissions.
- Automotive: CM help reduce vehicle weight, enhance fuel economy, and improve safety. Electric vehicles, in particular, benefit from their ability to extend battery range. Carbon fiber and glass fiber-reinforced plastics are common.
- Renewable Energy: CM are critical in wind turbine blades and solar energy components. Fiberglass composites offer the strength and lightweight profile needed for efficient energy capture and weather durability (Egbo, 2021; Chu et al., 2018).

Figure 9.4 illustrates major CM applications across these key sectors.

Figure 9.4: Applications span of CM in Industries.

Composite materials (CM) are widely adopted across sectors for their lightweight, strength, and corrosion resistance. In construction, fiber-reinforced polymer (FRP) composites are used in roofs, facades, bridges, and reinforcements —ideal for harsh environments like coastal and chemical sites. The marine sector benefits from CM in ship hulls and decks, reducing maintenance and improving fuel efficiency. In defense, CM are critical for lightweight vehicle parts, ballistic armor, and protective gear due to their strength and impact resistance. Sports industries use carbon fiber composites in equipment like bikes, rackets, and helmets for their strength-to-weight advantage. In electronics, CM serve in circuit boards and housings for their insulation and durability. Telecom relies on CM for lightweight, corrosion-resistant antenna structures. In healthcare, CM are employed in prosthetics and surgical tools for their biocompatibility and comfort. Finally, in oil and gas, CM are used in storage tanks and pipelines for their durability under extreme weather and chemicals (Das et al., 2020). Overall, CM are reshaping modern industries with high-performance capabilities, especially as automation continues to enhance their manufacturing precision and scalability.

9.3 Evolution of Automation in Manufacturing

9.3.1 Historical Background of Automation in Industry

Technological progress in manufacturing automation—from early mechanical systems to today's AI- and IoT-driven smart factories—has consistently aimed at enhancing productivity, precision, and efficiency. This evolution has played a pivotal role in advancing composite material production (Zhong et al., 2017).

9.3.1.1 Early Mechanization (18th to 19th Century)

Industrial automation began during the late 18th-century Industrial Revolution, when steam-powered machines replaced manual labor in industries like mining and textiles. Key inventions—the steam engine, power loom, and spinning jenny—laid the foundation for mechanized manufacturing by reducing reliance on human effort.

9.3.1.2 Electrification and Mass Production (Late 19th to Early 20th Century)

The electrification era transformed manufacturing by enabling more reliable and efficient production methods. It allowed the introduction of powered machinery and assembly lines, greatly increasing productivity. A major breakthrough came in 1913 when Henry Ford implemented the moving assembly line, revolutionizing mass production in the automotive sector by reducing labor needs and production time.

9.3.1.3 Automation and Control Systems (Mid-20th Century)

The mid-1900s saw major advances in electronics and control systems that sparked industrial automation. In the late 1960s, programmable logic controllers (PLCs) revolutionized factory operations by automating machinery and reducing human intervention. Automation quickly expanded across manufacturing, chemical, and automotive sectors. Robotics also emerged, with General Motors deploying the first industrial robot, Unimate, in 1961—a pivotal step in using robots for repetitive and hazardous tasks, significantly boosting productivity and workplace safety.

9.3.1.4 Digital Revolution and Computer-Aided Manufacturing (1970s to 1990s)

The rise of computer and digital technologies in the 1970s marked a major leap in automation. Computer-aided design and manufacturing enabled precise and efficient part production. Flexible manufacturing systems and just-in-time methods gained traction, helping industries adapt quickly to shifting demands. Integration of robotics, sensors, and computerized controls further boosted flexibility and productivity.

9.3.1.5 The Rise of Advanced Manufacturing and Industry 4.0 (2000s to Present)

Industry 4.0 and advanced manufacturing mark a shift toward interconnected, data-driven production using IoT, AI, and big data analytics. Intelligent systems now enable real-time communication, decision-making, and predictive maintenance. Digital twins (DT) further support process simulation and optimization, reducing errors and downtime. These innovations have accelerated automation in composite manufacturing, enhancing precision and efficiency through robotic cutting, automated fiber placement, and resin infusion (Bahrin et al., 2016).

9.3.1.6 Current Trends and Future Prospects

Automation continues to evolve with deeper integration of edge computing, AI, and ML, enabling highly responsive, efficient production through smart sensors, cobots, and autonomous systems. Sustainability is also shaping automation strategies, driving waste and energy reduction. From early mechanization to today's smart factories, automation remains central to industrial innovation—especially in composite material (CM) manufacturing, which exemplifies its transformative and scalable potential (Zheng et al., 2018).

9.3.2 The Rise of Smart Manufacturing Based on Industry 4.0

Industry 4.0 represents the integration of digital technologies—such as IoT, AI, and CPS—into manufacturing to enable smarter, data-driven operations. Building on past industrial revolutions, it introduces intelligent systems that improve flexibility, precision, and efficiency in modern factories (Morgan et al., 2021). IoT connects machinery and sensors for real-time data exchange, while CPS autonomously monitor and adapt to production conditions, reducing human intervention. AI and ML enhance predictive maintenance and process optimization by analyzing large datasets. Digital twin (DT) technology allows manufacturers to simulate and refine workflows before implementation (Huang et al., 2021; Di Pasquale et al., 2022). Additionally, cobots, edge/cloud computing, and advanced robotics support responsive and high-quality production. In composite manufacturing, Industry 4.0 boosts efficiency through AFP, robotic cutting, and AI-driven quality control (Ahuett-Garza and Kurfess, 2018). Despite challenges—like high costs, cybersecurity risks, and skill gaps—the long-term gains in sustainability, precision, and competitiveness make it a transformative approach. With continued adoption, CM producers can meet evolving industrial demands with greater agility and performance.

9.4 Automation in Composite Material Manufacturing

9.4.1 Key Processes in Composite Manufacturing

Automation has revolutionized composite material (CM) manufacturing by boosting precision, consistency, and efficiency across multi-step processes. Essential for aerospace, automotive, and renewable energy sectors, automation reduces material waste, labor, and production time.

Key technologies include:

- Automated Fiber Placement (AFP) and Automated Tape Laying (ATL): These systems ensure accurate fiber or tape deposition for structurally sound components, reducing defects in complex parts (Yassin and Hojjati, 2018).
- Vacuum-Assisted Resin Infusion (VARI) and Resin Transfer Molding (RTM): Automated resin infusion ensures uniform distribution and curing, minimizing manual intervention and enhancing quality (Rakhshbahar and Sinapius, 2018).
- Prepreg Processing: Automation improves precision in cutting and placement of pre-impregnated reinforcements, commonly used in high-performance applications (Kappel et al., 2013).
- Curing: Autoclaves with automated controls monitor time, temperature, and pressure in real time, enhancing uniformity and energy efficiency.
- Trimming and Machining: Robotic systems with sensors ensure accurate shaping and surface finish, reducing rework (Budelmann et al., 2020).
- Inspection and Quality Control: Automated systems using machine vision and ultrasonic NDT detect defects such as delamination and porosity, ensuring consistent product quality.

Overall, automation is streamlining CM production workflows, improving repeatability, and supporting rising demands for high-performance components.

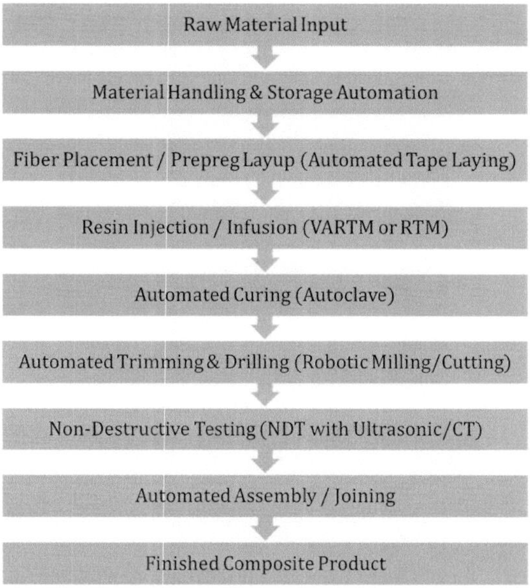

Figure 9.5: Flowchart of automated composite production process.

9.4.1.1 Automated Fiber Placement and Automated Tape Laying

Automated Fiber Placement (AFP) and Automated Tape Laying (ATL) are transformative technologies in composite manufacturing. These methods precisely deposit continuous fibers or prepreg tapes onto molds using robotic systems, enabling lightweight, high-strength components (Budelmann et al., 2020). Widely used in aerospace, automotive, and wind energy sectors, they enhance production speed, scalability, and accuracy. In AFP, resin-impregnated fibers are laid layer by layer with programmable control over fiber orientation, ensuring optimal structural performance and minimal defects. This is especially effective for large, curved, or complex parts like aircraft fuselages and wings, where precise alignment is critical. Figure 9.6 illustrates the AFP process managing complex 3D geometries with individual tows.

Automated Tape Laying (ATL) is designed to place wider composite tapes — ranging from a few centimeters to several inches—efficiently over large, flat, or mildly curved surfaces like wind turbine blades, automotive panels, and aircraft wings (Rakhshbahar and Sinapius, 2018). ATL offers high-speed layup, ideal for high-volume production by covering broad areas rapidly.Compared to manual methods, ATL and AFP deliver more consistent fiber placement, reducing defects, material waste, and labor demands. These automated systems allow complex layup paths and fiber orientations using advanced software, while real-time monitoring ensures quality control. Despite their advantages, high initial costs and the need for skilled operators and specialized equipment can limit adoption

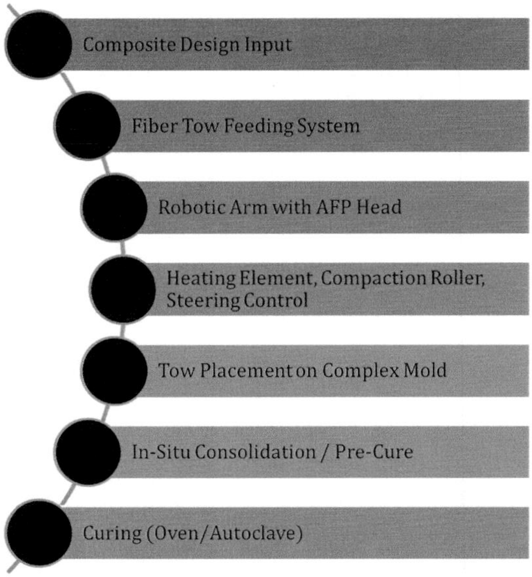

Figure 9.6: Step by step procedure of automated fiber placement process.

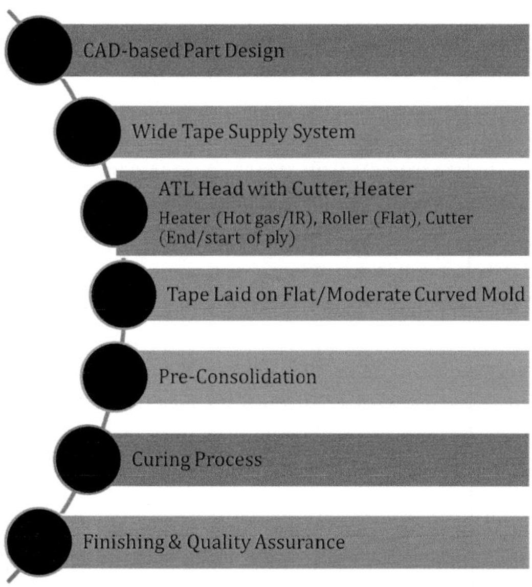

Figure 9.7: Step by step procedure of automated tape laying process.

(Bahar et al., 2022; Lukaszewicz et al., 2012). Nonetheless, their ability to improve scalability, speed, and structural reliability makes them central to advanced composite manufacturing.

Although AFP and ATL systems require complex programming and technical expertise, their long-term benefits—high accuracy, productivity, and efficient material use—make them crucial for modern composite manufacturing. As demand for lightweight, high-performance components grows, these technologies are expected to see wider adoption. Their integration with digital twins (DT), AI, and ML will drive smarter, more innovative, and efficient production capabilities (Budelmann et al., 2020).

9.4.1.2 Automation in Cutting, Forming, and Curing

Automation in cutting, shaping, and curing has become vital in composite material (CM) manufacturing, significantly enhancing precision, consistency, and throughput. Advanced systems—such as robotic cutters, water jets, and lasers—enable accurate shaping of complex laminates while minimizing material waste (Altintas and Ber, 2001). Sensor-based feedback ensures optimal material preparation and stacking.Automated forming using robotic arms and advanced heating techniques allows the production of intricate 3D structures with high repeatability, supporting applications where precision and quality are critical (Huang et al., 2023). Curing processes have also advanced through automation. Autoclaves with precise control systems and out-of-autoclave (OOA) ovens—like microwave-assisted curing—offer energy-efficient, uniform curing, essential for optimal mechanical properties. Real-time data and analytics further improve quality control by enabling adaptive adjustments (Aramburu and Vallejo, 2022). While high setup costs and skilled labor demands pose challenges, the long-term gains in efficiency, quality, and sustainability justify the investment. Future integration of AI, ML, and data-driven control is expected to further refine these processes, supporting scalable, eco-efficient CM production (Rong et al., 2022).

9.4.1.3 Inspection and Quality Control Automation

Automation is now essential for inspection and quality control in composite material (CM) manufacturing, ensuring products meet strict performance standards. Traditional manual checks struggle with detecting small defects in complex geometries, whereas automated systems enhance speed, accuracy, and consistency using machine vision and data analytics (Villalba-Diez et al., 2019). These systems identify surface flaws like delamination, fiber misalignment, or resin pooling with high precision. Non-destructive testing (NDT) has also advanced through automation, with robotic arms using ultrasonic, laser shearography, and infrared thermography to uncover internal defects efficiently (Ahmed et al., 2023). Coupled with AI, these systems analyze inspection data for predictive quality control, allowing early intervention and ongoing process optimization.Digital twins (DT) further strengthen quality oversight by comparing real-time production data with virtual models to detect anomalies and monitor component health (Hoffmann and Reich, 2023). Although initial costs and skilled labor needs pose challenges, the long-term gains—reduced inspection time, fewer defects, and better product consistency—make automation a valuable investment. As demand for high-quality CM rises, future advances in AI, ML, and real-time analytics are expected to further refine inspection accuracy and efficiency.

9.4.2 Role of Robotics in Composite Manufacturing

Robotics has significantly transformed composite material (CM) manufacturing by enhancing output, accuracy, and consistency—ideal for this labor-intensive process. As demand for lightweight, high-performance components grows in sectors like wind energy, automotive, and aerospace, robotics helps streamline production, reduce costs, and maintain quality.

Key contributions include:

- Automated layup (AFP & ATL): Robots accurately align and place fibers or prepreg tapes, improving part uniformity and handling complex geometries (Parmar et al., 2022).
- Material handling and forming: Robotic arms reduce errors during shaping and ensure gentle, precise manipulation, minimizing damage and accelerating throughput (İpekçi and Ekici, 2024).
- Cutting and machining: Robots equipped with water jets, lasers, and milling heads achieve high-precision trimming and adaptable cutting paths.
- Curing support: Robotic systems manage loading/unloading and maintain optimal curing parameters, boosting integrity and repeatability.
- Inspection (NDT): Robots with sensors and vision systems detect internal and surface defects faster and more reliably, improving quality control.

While initial costs and specialized training are barriers (Gupta et al., 2023), robotics offers long-term gains in capacity, quality, and material efficiency. Future advances—including cobots, AI, and ML—promise greater adaptability and smart human-robot collaboration (Le and Tran, 2024; Frketic et al., 2017), keeping robotics central to next-gen CM manufacturing.

9.4.3 Role of AI and ML in Automation

Machine learning (ML) and artificial intelligence (AI) are transforming composite material (CM) manufacturing by enabling process optimization, predictive maintenance, optimization, and data-driven decision-making. These technologies analyze large datasets from production to improve quality, reduce waste, and enhance efficiency (Peng and Bhaskar, 2023). AI systems dynamically optimize parameters like temperature and material feed rates, minimizing defects and boosting output. For maintenance, ML models detect early signs of

Table 9.1: Summary of research gaps and opportunities for further developments in robotics for composite manufacturing.

S. No.	Category	Current challenges	Opportunities
1.	Path Planning & Kinematics	Inflexible trajectories, poor real-time response	ML-based dynamic path planning with adaptive control
2.	Quality Control	Post-process inspection leads to rework	Integrate NDT and vision systems for inline QC
3.	End-Effector Design	Limited adaptability for complex geometries	Develop modular and smart end-effectors
4.	Collaborative Robotics	Safety concerns and payload limitations	Enhance cobot strength and situational awareness
5.	Digital Twin & AI	High computational cost and deployment issues	Use edge-AI and simplified twin architectures
6.	Sustainability	Material waste and energy use in robotic processes	Design energy-efficient, recyclable systems with LCA focus

equipment failure using real-time sensor data, reducing downtime and operational costs. In quality control, AI-driven machine vision identifies defects such as fiber misalignment or delamination more accurately and rapidly than manual inspections (Sacco et al., 2020). AI also helps trace defect origins, supporting continuous improvement (Woschank et al., 2020; Sikka et al., 2024). Digital twins (DT), powered by AI analytics, create virtual replicas of manufacturing processes, allowing simulation, monitoring, and optimization of operations in real time. Additionally, AI enhances robotic systems—particularly collaborative robots (cobots)—which learn from human operators to improve task execution and flexibility. While challenges like data security and system maintenance exist, the long-term benefits of AI and ML include improved product consistency, lower costs, and scalable smart manufacturing. Integrated with Industry 4.0 technologies, they position CM manufacturers to stay competitive and innovative (Wang and Siau, 2019; Sacco et al., 2020).

9.4.4 Human-Machine Collaboration

Human–machine collaboration is transforming composite material (CM) manufacturing by combining human creativity with the precision of automation. As automation technologies like robotics, AI, and ML advance, human roles are shifting from manual tasks to supervising and refining automated processes, enhancing flexibility, safety, and productivity (Bryksin et al., 2023). Humans excel at addressing complex, non-standard challenges where automated systems struggle. Operators can adjust parameters and resolve unexpected issues, ensuring responsiveness to diverse production needs. Collaborative robots (cobots), guided by AI, safely assist with tasks like material handling, assembly, and inspection, reducing worker fatigue and injury. AI systems also support decision-making by analyzing real-time data and alerting operators to defects or equipment failures. This synergy improves quality and reduces waste, but it demands skilled workers trained in automation technologies. Continuous training is essential as roles evolve toward system oversight and optimization. Challenges include ensuring clear communication between humans and machines, managing technological change, and addressing cybersecurity. Moving forward, integrating AI, digital twins (DT), and intuitive interfaces will foster smarter, more adaptive manufacturing ecosystems—driving innovation and competitiveness in CM production (Nevliudov et al., 2021; Zhang and Kwok, 2018).

9.5 Benefits of Automation in Composite Manufacturing

Automation in composite manufacturing enhances productivity while reducing costs and material waste. Technologies like robotic fiber placement and automated resin infusion minimize defects and human error, ensuring consistent quality and structural performance. These systems also accelerate production to meet rising demand in aerospace, automotive, and renewable energy sectors (Arrabiyeh et al., 2021). Additionally, automation improves worker safety by limiting exposure to hazardous materials and reducing physically demanding tasks. Figure 9.8 highlights these key benefits.

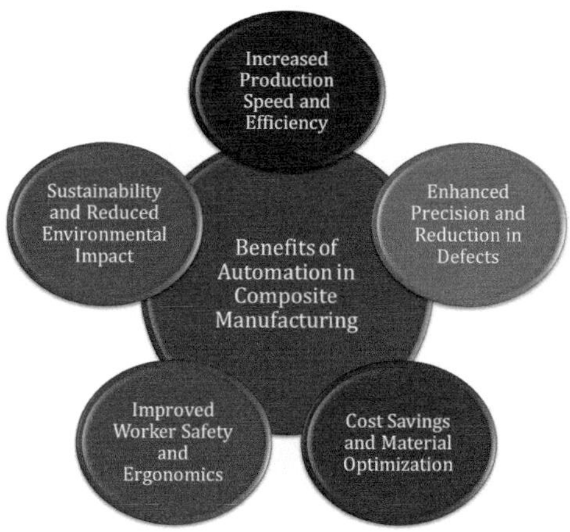

Figure 9.8: Benefits of automation in composite manufacturing

9.5.1 Enhanced Production Speed and Efficiency

Automation greatly boosts production speed and efficiency in composite manufacturing by replacing traditional labor-intensive methods like hand layup with advanced robotic systems and computer-controlled machinery. Technologies such as Automated Tape Laying (ATL) and Automated Fiber Placement (AFP)

enable precise, high-speed deposition of fibers in complex patterns, meeting high-volume demand without compromising quality (Sacco et al., 2020). Integrated cutting and trimming systems—using waterjet and laser technology—streamline shaping, while automated production lines ensure smoother transitions between layup, curing, and inspection. Machine learning (ML) and predictive analytics further enhance productivity by identifying performance patterns and enabling timely maintenance. Automation reduces labor needs for repetitive tasks, reallocates human effort to higher-value activities, and minimizes material waste—improving overall resource use and cost efficiency. These factors are key for maintaining competitiveness in industries like aerospace and automotive (Domm, 2020). As smart technologies evolve, they will continue to optimize production and support growing market demand.

9.5.2 Enhanced Precision and Reduction in Defects

Automation significantly enhances the accuracy and reliability of composite material (CM) manufacturing by minimizing human error and improving consistency through robotics, computer-controlled systems, and real-time monitoring. Automated processes like ATL and AFP allow for precise fiber placement and uniform material distribution, ensuring components meet strict design standards (Wang et al., 2017; Ismail et al., 2022). Sophisticated control systems dynamically adjust variables like temperature, pressure, and material flow, reducing waste and rework. Automated curing, particularly out-of-autoclave (OOA) techniques, ensures optimal thermal conditions, minimizing voids and defects while boosting product durability. Non-destructive testing (NDT) and machine vision technologies further strengthen quality assurance by accurately detecting flaws such as fiber misalignments and delamination. These insights also aid process improvement, reducing failures and cutting costs (Hoa, 2018; Loeliger et al., 2021). Though initial investments may be high, the long-term gains in quality, efficiency, and sustainability make automation indispensable for high-performance CM manufacturing.

9.5.3 Cost Savings and Material Optimization

Automation in composite material (CM) manufacturing enhances material efficiency and reduces costs, making it a strategic investment for competitiveness. Traditional manual methods often incur high labor costs, poor resource use, and excess waste. In contrast, automated systems—such as computer-controlled machinery and robotics—perform repetitive tasks with greater precision and less error, maintaining high output with reduced rework. Technologies like AFP and ATL ensure accurate fiber placement, minimizing material waste and improving structural performance. Real-time monitoring systems adjust processing parameters to optimize curing conditions and avoid defects (Chen et al., 2020).

Automation also supports just-in-time (JIT) manufacturing, reducing inventory and storage costs while boosting operational agility. Though initial investments are high, the long-term ROI is strong—thanks to increased scalability, consistent quality, and lower warranty claims (Chen et al., 2022). As automation technologies evolve, further gains in cost savings and material optimization are expected, cementing their role in next-generation CM production.

9.5.4 Improved Worker Safety and Ergonomics

Automation significantly enhances worker safety and ergonomics in composite material (CM) manufacturing by minimizing exposure to hazardous tasks and improving overall workplace conditions. Traditional processes often involve heavy lifting, toxic chemicals, and high-temperature environments. Automated systems—like robotic arms and collaborative robots (cobots)—reduce physical strain and musculoskeletal risks by handling repetitive or dangerous jobs (Lowe et al., 2023). Automation also limits worker contact with harmful solvents and high-pressure curing setups, fostering a safer and healthier environment. Real-time monitoring and sensor-based controls further enhance safety by detecting system faults or hazards early (Gualtieri et al., 2020). Cobots not only assist in precision tasks but allow human workers to focus on more cognitively engaging roles, increasing job satisfaction. Additionally, automation supports consistent compliance with safety standards, lowering workplace accidents and insurance costs (Colim et al., 2021). While initial investments and training are required, the long-term benefits—reduced injuries, improved well-being, and higher operational safety—make automation a valuable asset in modern CM production.

9.5.5 Sustainability and Reduced Environmental Impact

Automation plays a pivotal role in advancing sustainability in composite material (CM) manufacturing by improving energy efficiency, reducing waste, and optimizing resource usage. Technologies like Automated Tape Laying (ATL) and Automated Fiber Placement (AFP) enhance material utilization and minimize scrap through precise deposition, reducing the need for post-processing (Ford and Despeisse, 2016). Energy-intensive steps like curing are streamlined by automation with optimized control systems and real-time monitoring, while out-of-autoclave (OOA) methods further cut energy use and emissions (Dhiman et al., 2021). Data analytics supports energy planning by identifying inefficiencies and enabling off-peak scheduling. Automation also enhances recyclability by producing fewer flawed parts and supporting bio-based material integration—facilitating circular economy goals. Emissions are reduced through accurate handling of volatile compounds and cleaner processes managed by AI and robotics. Aligned with Industry 4.0, digital tools like IoT, AI, and digital twins (DT) support predictive maintenance and environmental tracking, promoting a greener industrial ecosystem. Despite upfront costs, automation offers long-term environmental and economic benefits, with evolving technologies poised to further strengthen its sustainability impact (Witik et al., 2012).

9.6 Industry Case Studies

9.6.1 Automation in Aerospace Composite Manufacturing

The aerospace industry has led the adoption of automation in composite material (CM) manufacturing to meet its rigorous standards for safety, performance, and precision. Technologies like Automated Fiber Placement (AFP) and Automated Tape Laying (ATL) have been widely implemented by companies such as Boeing and Airbus to fabricate complex fuselages and wing structures with minimal waste and consistent quality (Dell'Anno et al., 2012). These systems precisely lay carbon fibers on large molds, enhancing structural integrity and speeding up production—key to developing aircraft like the Airbus A350 XWB and Boeing 787. Beyond layup, robotics are extensively used in assembling, drilling, and fastening, ensuring alignment accuracy and reducing defects in composite components (Chen et al., 2023).

In aerospace composite manufacturing, automation enhances precision and structural integrity—especially in drilling and fastening—through advanced sensors and robotic end-effectors. This ensures tight alignment and reduces defects, as seen in Lockheed Martin's F-35 program (Sacco et al., 2020; Blake, 2019). Automated curing systems, often with real-time monitoring, optimize thermal cycles while reducing energy use. The adoption of out-of-autoclave (OOA) techniques furthers sustainability. Automated inspection using machine vision and ultrasonic testing boosts flaw detection speed and accuracy, critical for safety compliance. NASA's use of robotic NDT exemplifies these advancements. Beyond improving quality and throughput, automation reduces emissions, waste, and energy consumption. Looking ahead, deeper integration of digital twins (DT), machine learning (ML), and AI will enable real-time optimization and predictive maintenance, keeping the aerospace sector at the forefront of smart manufacturing (Eugeni et al., 2022). Let me know if you'd like a flowchart or case summary to visualize these elements.

9.6.2 Automotive Industry's Use of Automated Composite Production

Automation is transforming composite material (CM) manufacturing in the automotive sector, driven by the demand for strong, lightweight parts to improve fuel efficiency and sustainability. Manufacturers like BMW and Tesla have implemented fully automated systems—including robotic layup, curing, cutting, and assembly—to scale production of carbon fiber reinforced polymers (CFRP) while maintaining structural integrity (Sarfraz et al., 2021). BMW's automated line for the i3/i8 and Audi's use of automated resin transfer molding (RTM) for

A8 roof components exemplify how automation shortens cycle times and enhances quality. Robotic assembly ensures precise drilling and bonding, while automated non-destructive testing (NDT), including thermographic and ultrasonic methods, improves defect detection and reduces waste. As EV and autonomous vehicle production grows, automation enables scalable, cost-effective manufacturing of lightweight CM parts. Future integration of AI, ML, and predictive analytics is expected to further optimize design, inspection, and production processes (Sacco et al., 2020; Blake, 2019), keeping the automotive industry competitive and innovative.

9.6.3 Renewable Energy and Automation

Automation in composite manufacturing has significantly advanced wind energy, especially in producing larger, high-performance turbine blades. These blades require durable, lightweight materials like fiberglass and carbon fiber, which are precisely placed using Automated Tape Laying (ATL) and Automated Fiber Placement (AFP) technologies. Such systems enhance blade alignment, strength, and uniformity while reducing errors and production time (Nejad et al., 2022; Dimitrova et al., 2022). Leading firms like Siemens Gamesa and Vestas use robotics for molding and automated resin infusion, ensuring accurate material distribution and minimal defects. Automated curing systems—often integrated with real-time monitoring and Out-of-Autoclave (OOA) techniques—further support energy-efficient, consistent processing, promoting sustainability and cost savings (Komarov et al., 2011).

Automation has transformed wind turbine blade manufacturing by enhancing defect detection and precision. Laser scanning and ultrasonic inspection—key automated non-destructive testing (NDT) methods—accurately identify internal and surface flaws, enabling real-time feedback and quality assurance. AI-powered machine vision further boosts detection speed and accuracy. Automation also enables the efficient production of longer, more complex blades (100+ meters), which are critical for offshore wind farms. This not only enhances energy output but improves worker safety by reducing manual handling of large components. Aligned with sustainability goals, automated systems minimize material waste, optimize energy use, and reduce emissions, making blade production more environmentally responsible (Martinez-Marquez et al., 2022). As automation evolves, integrating AI, ML, and predictive analytics will enhance efficiency and adaptability, further supporting the growth of renewable energy (La Cava et al., 2016).

9.6.4 Innovations in the Marine and Construction Sectors

Automation is increasingly adopted in construction and maritime industries to meet demands for stronger, lightweight, and corrosion-resistant composite materials (CM) like carbon fiber and fiberglass. Traditional production methods struggle with scalability, whereas automated technologies improve efficiency, precision, and innovation. In the maritime sector, companies such as Austal and Damen Shipyards employ Automated Tape Laying (ATL), robotic filament winding, and automated resin transfer molding (RTM) to streamline hull and deck production with enhanced material deposition and reduced defects (Doloreux and Melançon, 2009). These processes are especially beneficial for high-performance vessels. In construction, automated pultrusion ensures dimensional consistency in bridge and façade components, while additive manufacturing enables complex, lightweight structures for modular building. 3D printing of CM expands design possibilities not achievable with conventional methods. AI-powered inspection and automated non-destructive testing (NDT) tools improve safety and reduce rework through real-time defect detection (Su et al., 2022). These advancements also support sustainability by minimizing waste and carbon emissions. Overall, automation boosts quality, productivity, and design flexibility—key for maintaining competitiveness across sectors. Continued innovation will further enhance efficiency and environmental performance in CM applications (Wiśnicki et al., 2024).

9.7 Challenges and Limitations of Automation

9.7.1 Technical Challenges: Complexity and Integration

Despite its transformative impact,automation in composite material (CM) manufacturing faces technical hurdles—particularly around system integration and process complexity. Automating fiber placement, resin infusion, curing, and inspection demands precise control and advanced coordination of robotics, sensors, and real-time monitoring.

Key challenges include:

- Material variability: CMs are anisotropic, making fiber orientation, resin content, and curing conditions harder to standardize. Even minor misalignments can impact structural integrity (Ahmed et al., 2023).
- Legacy system compatibility: Many facilities still rely on manual or semi-automated workflows, complicating integration and requiring software updates and workforce retraining.
- Sensor and algorithm demands: Ensuring quality control across diverse part geometries needs sophisticated monitoring systems and adaptive AI.
- Robotic handling of flexible materials: The delicate nature of CMs requires precise actuation to prevent defects.

Emerging solutions—including digital twins (DT), machine learning (ML), and smart monitoring—are addressing these integration challenges (Villalba-Diez et al., 2019). Continued investment in infrastructure, skilled personnel, and R&D is essential to fully realize automation's potential in CM production.

9.7.2 Economic and Investment Barriers

Automation in composite material (CM) manufacturing presents financial challenges, particularly for small and medium-sized enterprises (SMEs). High initial costs—for robotics, AI-enabled monitoring, and system integration—often slow full-scale adoption. While these technologies boost productivity and accuracy, their return on investment (ROI) may be delayed due to added expenses for software, maintenance, and skilled labor (Fuchs et al., 2008). The shortage of trained personnel in robotics and data analytics also raises recruitment and training costs. Economic fluctuations and regulatory shifts in sectors like aerospace and renewable energy further affect automation investments. Despite these hurdles, growing support through government incentives and advances in AI and robotics are making automation more accessible. Strategic investments in automation—supported by partnerships and workforce development—offer long-term gains in efficiency, sustainability, and quality for CM manufacturing (Su et al., 2022).

9.7.3 Workforce Challenges: Skills, Training, and Job Displacement

Automation in composite material (CM) manufacturing presents workforce challenges, particularly in training, upskilling, and job displacement. As robotics, AI, and ML advance, demand for skilled personnel proficient in digital tools and smart systems grows. However, many current workers lack experience with automated technologies, and training programs—though essential—are often costly and time-intensive (Basualdo et al., 2021). Job displacement is also a concern, as automation reduces the need for manual labor in repetitive tasks. Yet, it creates new roles in areas like AI-based process control, data analytics, and robotics maintenance—requiring workforce reskilling (Howcroft and Taylor, 2023). Regional disparities in skilled labor further complicate adoption. To address these challenges, collaboration between industry and academia is key. Developing robust training initiatives, apprenticeships, and open communication can ease the transition and strengthen human-machine integration—ensuring long-term success in automated CM production.

9.7.4 Cyber Security and Data Privacy Concerns

Automation in composite manufacturing increasingly relies on digital technologies, heightening risks around data privacy and cybersecurity. Smart factories using IoT, cloud computing, and real-time monitoring are vulnerable to cyber threats like ransomware, data breaches, and system disruptions—jeopardizing production and IP integrity (Shackelford, 2018). Industrial control systems (ICS), such as PLCs and robotic networks, are especially susceptible. Attacks can lead to material defects or safety risks. IP theft is a particular concern in aerospace and defense, where proprietary data may be compromised through weak encryption or remote access (Ani et al., 2017). AI-driven automation introduces additional risks if training datasets are corrupted, potentially leading to faulty outputs. Global supply chains further amplify vulnerabilities due to outdated protocols and inconsistent security standards. To mitigate these threats, manufacturers must adopt layered cybersecurity strategies—encryption, access controls, timely software updates, and employee training—while aligning with regulatory standards like GDPR. A robust cybersecurity framework is essential to safeguard reliability and operational continuity in automated CM production (Mullet et al., 2021).

Table 9.2: Automation in composite manufacturing: Benefits vs. Challenges

Category	Benefits of automation	Challenges/Limitations
Productivity	- Faster production cycles - Higher throughput and efficiency - Consistent process speeds	- High initial setup and equipment cost - Downtime during integration and upgrades
Quality & Accuracy	- Improved repeatability and precision - Minimization of human error - Uniform material layup and curing	- Difficulty in handling complex geometries or contours - Defect correction often not real-time
Labor & Safety	- Reduced need for manual labor - Minimizes operator fatigue and injuries - Supports remote operation and monitoring	- Requires skilled technicians for programming and maintenance - Potential resistance to job displacement
Process Control	- Real-time monitoring and control (with sensors/AI) - Adaptive manufacturing using ML algorithms - Data-driven process optimization	- Complex integration of sensor data and AI - High data processing and storage needs
Cost Efficiency	- Lower operational costs over time - Reduced waste and rework - Improved material utilization	- High capital investment and maintenance cost - Long ROI period for small-scale operations
Flexibility & Scalability	- Reprogrammable for different part types - Compatible with modular manufacturing setups - Enables hybrid techniques (e.g., AFP + 3D printing)	- Reconfiguration may require downtime - Limited flexibility in legacy systems
Inspection & Quality Assurance	- Inline NDT (ultrasound, vision) integration - Automated defect detection and classification	- Inline NDT is still not mature for complex structures - False positives/negatives in detection systems
Sustainability	- Optimized use of raw materials - Lower carbon footprint through efficient processes	- Energy-intensive systems without green upgrades - Sustainability metrics often not integrated into automation tools
Digital Integration	- Compatibility with Industry 4.0 technologies - Supports digital twins, predictive maintenance	- Digital twin models are complex to build and validate - Cybersecurity and data integrity concerns

9.8 Future Trends in Automation for Composite Manufacturing

9.8.1 Emerging Technologies: 3D Printing, Additive Manufacturing

3D printing and additive manufacturing are reshaping composite material (CM) production by improving efficiency, reducing waste, and enabling complex, customizable designs. These layer-by-layer techniques reinforce polymer, metal, or ceramic matrices with fibers such as carbon or glass, producing high-performance structures with minimal material loss—ideal for aerospace, automotive, and renewable energy applications (Praveena et al., 2022). Large-scale additive manufacturing (LSAM) accelerates fabrication of components like turbine blades and fuselages, cutting lead times and cost. Hybrid systems that combine printing, machining, and real-time inspection enhance precision and reduce post-processing needs. Digital twin (DT) and AI simulations optimize process parameters and error prediction. While issues like material limits and scalability persist, advances in automation, AI, and process control are addressing these challenges. As technologies mature, 3D printing will play a key role in developing next-gen, high-performance composites with reduced environmental impact (Bin Wang et al., 2020).

9.8.2 Advancements in Robotics and AI in Composite Production

The integration of robotics and AI has revolutionized composite material (CM) manufacturing by enhancing automation, precision, and efficiency. Industrial production is increasingly shaped by intelligent automation, predictive analytics, and autonomous decision-making. Automated Fiber Placement (AFP) and Automated Tape Laying (ATL) streamline fiber alignment and curing, outperforming labor-intensive manual methods through higher throughput and reduced waste. Robotic arms ensure consistent layer deposition, improving mechanical properties (Parmar et al., 2022). AI-driven monitoring and predictive maintenance optimize CM output by detecting material anomalies and environmental deviations via sensor data. Machine learning (ML) models dynamically adjust fiber orientation, resin flow, and curing parameters to enhance quality (Gupta et al., 2023). Collaborative robots (cobots) further aid in material handling and inspection, operating safely alongside humans for greater flexibility. AI-powered digital twin (DT) systems enable engineers to simulate and optimize manufacturing workflows, minimizing errors and boosting efficiency. Despite challenges like system integration, high costs, and workforce training, growing research and affordability are accelerating adoption. Ultimately, robotics and AI are redefining CM manufacturing by improving productivity, scalability, and innovation (Priyadharshini et al., 2022).

9.8.3 The Role of DT and Simulation

The integration of digital twin (DT) and simulation technologies is transforming composite manufacturing (CM) by enabling real-time monitoring, predictive analysis, and process optimization. These tools allow virtual replication of manufacturing workflows, boosting productivity, minimizing errors, and accelerating the development of high-performance composites. Using sensor data and IoT devices, DT technology synchronizes physical and digital systems to simulate scenarios, predict failures, and fine-tune parameters—cutting downtime and waste (Polini and Corrado, 2020). Simulation-driven automation, supported by CFD and FEA, improves pre-production evaluation of thermal and mechanical behavior. AI-enhanced simulations further refine fiber placement, material selection, and curing processes, reducing trial-and-error. DT-enabled quality control also addresses internal defects like voids and delamination more effectively than traditional methods (Phanden et al., 2020). By optimizing robotic trajectories and process parameters in AFP, ATL, and additive manufacturing, DT reduces human intervention and setup time. Despite computational and modeling challenges, advances in AI, cloud computing, and HPC are enhancing accessibility. As industries embrace smart manufacturing, DT will continue to drive efficiency, innovation, and sustainability (Hürkamp et al., 2020).

9.8.4 Future Impact on Global Manufacturing Ecosystems

Automation is reshaping global composite material (CM) production by enhancing supply chain efficiency, sustainability, and scalability. Emerging technologies—AI, robotics, digital twin (DT), and additive manufacturing—are driving smart, autonomous systems that reduce labor dependency and enable consistent, cost-effective output. Data-driven workflows are replacing traditional manual processes, accelerating adoption in sectors like wind energy, automotive, and aerospace. Industry 4.0 tools support demand forecasting, inventory optimization, and real-time coordination across supply chains (Das and Dey, 2021). Cloud platforms further decentralize operations, enabling agile, remote-controlled production. Sustainability is also advancing, with AI-based predictive maintenance reducing energy consumption and precision processes like AFP minimizing material waste (Sautter, 2021). As low-skill roles decline, workforce demands are shifting toward expertise in robotics, AI, and DT modeling—necessitating upskilling initiatives. While disparities in automation adoption may widen global economic gaps, collaboration and technology sharing can help bridge them. Ultimately, automation is revolutionizing CM manufacturing, with digital connectivity, skilled labor development, and strategic investment shaping its future (Hughes et al., 2022).

9.9 Conclusion

Automation has transformed composite material (CM) manufacturing by significantly enhancing productivity, precision, and scalability across industries. Technologies such as digital twins (DT), robotics, AI, and additive manufacturing have replaced traditional labor-intensive methods, enabling higher throughput, reduced material waste, and superior quality. Systems like AFP and ATL accelerate fabrication of complex composites with improved structural integrity. Automation also optimizes resource use—minimizing energy consumption and emissions—while AI-powered predictive maintenance reduces downtime. These advancements support sustainable practices and long-term cost savings. Smart factories and integrated digital supply chains, powered by DT and Industry 4.0, improve flexibility and real-time coordination, fostering a more resilient industrial ecosystem. Looking ahead, automation is key to developing affordable, high-performance composites for sectors such as construction, automotive, aerospace, and renewable energy. As automation becomes essential, challenges like workforce upskilling, cybersecurity, and high initial investments must be addressed. With strategic planning and technological progress, automated CM production will continue to evolve—driving innovation, sustainability, and next-generation manufacturing (Chen et al., 2020).

References

Ahmed, Md Wadud, Sahir Junaid Hossainy, Alin Khaliduzzaman, Jason Lee Emmert, Mohammed Kamruzzaman et al. 2023. Non-destructive optical sensing technologies for advancing the egg industry toward Industry 4.0: A review. *Comprehensive Reviews in Food Science and Food Safety*. https://doi.org/10.1111/1541-4337.13227.

Ahuett-Garza, H. and Kurfess, T. 2018. A brief discussion on the trends of habilitating technologies for Industry 4.0 and smart manufacturing. *Manufacturing Letters*. https://doi.org/10.1016/j.mfglet.2018.02.011.

Ali, Alamry and Andri Andriyana. 2020. Properties of multifunctional composite materials based on nanomaterials: A review. *RSC Advances*. https://doi.org/10.1039/c9ra10594h.

Altintas, Y. and Ber, A.A. 2001. Manufacturing automation: Metal cutting mechanics, machine tool vibrations, and CNC design. *Applied Mechanics Reviews*. https://doi.org/10.1115/1.1399383.

Alves de Campos, António, Elsa Henriques and Christopher L. Magee. 2022. Technological improvement rates and recent innovation trajectories in automated advanced composites manufacturing technologies: A patent-based analysis. *Composites Part B: Engineering*. https://doi.org/10.1016/j.compositesb.2022.109888.

Ani, Uchenna P. Daniel, Hongmei (Mary) He and Ashutosh Tiwari. 2017. Review of cybersecurity issues in industrial critical infrastructure: Manufacturing in perspective. *Journal of Cyber Security Technology*. https://doi.org/10.1080/23742917.2016.1252211.

Aramburu, A. and Vallejo, F.J. 2022. Development and automation of a new composites manufacturing process based on UV curing prepregs. *Revista de Materiales Compuestos*. https://doi.org/10.23967/r.matcomp.2022.03.004.

Arrabiyeh, Peter A., David May, Maximilian Eckrich, Anna M. Dlugaj et al. 2021. An overview on current manufacturing technologies: Processing continuous rovings impregnated with thermoset resin. *Polymer Composites*. https://doi.org/10.1002/pc.26274.

Bahar, Mohammad, Marco Brysch and Michael Sinapius. 2022. *In situ* consolidation of thermoplastic prepreg by generating harmonic oscillations on the consolidation roller. *Journal of Composites Science*. https://doi.org/10.3390/jcs6010001.

Bahrin, Mohd Aiman Kamarul, Mohd Fauzi Othman, Nor Hayati Nor Azli and Muhamad Farihin Talib. 2016. Industry 4.0: A review on industrial automation and robotic. *Jurnal Teknologi*. https://doi.org/10.11113/jt.v78.9285.

Basualdo, Victoria, Hugo Dias, Mirko Herberg, Stefan Schmalz, Melisa Serrano et al. 2021. Building workers' power in digital capitalism: Old and new labour struggles. *Old and New Labour ....*

Bhong, Mahesh, Tasneem K.H. Khan, Kiran Devade, B., Vijay Krishna, Sreekanth Sura et al. 2023. Review of composite materials and applications. *Materials Today: Proceedings*. https://doi.org/10.1016/j.matpr.2023.10.026.

Blake, Scott. 2019. Recent advances in artificial intelligence applications to composites fabrication. In *CAMX 2019—Composites and Advanced Materials Expo*. https://doi.org/10.33599/nasampe/c.19.0658.

Bryksin, Matvei, Nikita Vysotsky and Pavel Guseynov. 2023. Industrial HMI design principles for highly automated manufacturing processes. In *Human Interaction & Emerging Technologies (IHIET 2023): Artificial Intelligence & Future Applications*. https://doi.org/10.54941/ahfe1004043.

Budelmann, Dennis, Carsten Schmidt and Dieter Meiners. 2020. Prepreg tack: A review of mechanisms, measurement, and manufacturing implication. *Polymer Composites*. https://doi.org/10.1002/pc.25642.

Cava, William La, Kourosh Danai, Lee Spector, Paul Fleming, Alan Wright et al. 2016. Automatic identification of wind turbine models using evolutionary multiobjective optimization. *Renewable Energy*. https://doi.org/10.1016/j.renene.2015.09.068.

Chen, Shize, Daochun Li and Jinwu Xiang. 2022. A novel composite design optimization method for minimized manufacturing cost with improved performances. *Applied Composite Materials*. https://doi.org/10.1007/s10443-022-10027-2.

Chen, Shize, Daochun Li, Jinwu Xiang and Shiwei Zhao. 2020. Composite manufacturing cost model targeting on design optimization. *Applied Composite Materials*. https://doi.org/10.1007/s10443-020-09828-0.

Chen, Yonglin, Junming Zhang, Zefu Li, Huliang Zhang, Jiping Chen et al. 2023. Manufacturing technology of lightweight fiber-reinforced composite structures in aerospace: Current situation and toward intellectualization. *Aerospace*. https://doi.org/10.3390/aerospace10030206.

Choi, Jun Young, Ji Ho Jeon, Jang Hyeon Lyu, Jungwoo Park, Geun Young Kim et al. 2023. Current applications and development of composite manufacturing processes for future mobility. *International Journal of Precision Engineering and Manufacturing—Green Technology*. https://doi.org/10.1007/s40684-022-00483-3.

Chu, Zhaoqiang, Mohammadjavad Pourhosseiniasl and Shuxiang Dong. 2018. Review of multi-layered magnetoelectric composite materials and devices applications. *Journal of Physics D: Applied Physics*. https://doi.org/10.1088/1361-6463/aac29b.

Colim, Ana, Rita Morgado, Paula Carneiro, Nélson Costa, Carlos Faria et al. 2021. Lean manufacturing and ergonomics integration: Defining productivity and wellbeing indicators in a human–robot workstation. *Sustainability (Switzerland)*. https://doi.org/10.3390/su13041931.

Das, Arindam and Sourav Dey. 2021. Global manufacturing value networks: Assessing the critical roles of platform ecosystems and Industry 4.0. *Journal of Manufacturing Technology Management*. https://doi.org/10.1108/JMTM-04-2020-0161.

Das, Monalisa, Sasmita Sahu and Dayal R. Parhi. 2020. A review of application of composite materials for aerospace structures and its damage detection using artificial intelligence techniques. *SSRN Electronic Journal*. https://doi.org/10.2139/ssrn.3714181.

Dell'Anno, Giuseppe, Ivana Partridge, Denis Cartié, Alexandre Hamlyn, Edmon Chehura et al. 2012. Automated manufacture of 3D reinforced aerospace composite structures. *International Journal of Structural Integrity*. https://doi.org/10.1108/17579861211209975.

Dhiman, Sahil, Ravinder Singh Joshi, Sachin Singh, Simranpreet Singh Gill, Harpreet Singh et al. 2021. A framework for effective and clean conversion of machining waste into metal powder feedstock for additive manufacturing. *Cleaner Engineering and Technology*. https://doi.org/10.1016/j.clet.2021.100151.

Dimitrova, Mariya, Ahmad Aminzadeh, Mohammad Saleh Meiabadi, Sasan Sattarpanah Karganroudi, Hossein Taheri and Hussein Ibrahim. 2022. A survey on non-destructive smart inspection of wind turbine blades based on Industry 4.0 Strategy. *Applied Mechanics*. https://doi.org/10.3390/applmech3040075.

Doloreux, David and Yannik Melançon. 2009. Innovation-support organizations in the marine science and technology industry: The case of quebec's coastal region in Canada. *Marine Policy*. https://doi.org/10.1016/j.marpol.2008.04.005.

Domm, Matthias. 2020. Printing of three-dimensional polymer composite structures with continuous fiber reinforcement. In *Structure and Properties of Additive Manufactured Polymer Components*. https://doi.org/10.1016/B978-0-12-819535-2.00011-9.

Drissi, Sarra, Tung Chai Ling, Kim Hung Mo and Anissa Eddhahak. 2019. A review of microencapsulated and composite phase change materials: Alteration of strength and thermal properties of cement-based materials. *Renewable and Sustainable Energy Reviews*. https://doi.org/10.1016/j.rser.2019.04.072.

Egbo, Munonyedi Kelvin. 2021. A fundamental review on composite materials and some of their applications in biomedical engineering. *Journal of King Saud University—Engineering Sciences*. https://doi.org/10.1016/j.jksues.2020.07.007.

Eugeni, M., Quercia, T., Bernabei, M., Boschetto, A., Costantino, F. et al. 2022. An Industry 4.0 approach to large scale production of satellite constellations. the case study of composite sandwich panel manufacturing. *Acta Astronautica*. https://doi.org/10.1016/j.actaastro.2021.12.039.

Ford, Simon and Mélanie Despeisse. 2016. Additive manufacturing and sustainability: An exploratory study of the advantages and challenges. *Journal of Cleaner Production*. https://doi.org/10.1016/j.jclepro.2016.04.150.

Frketic, Jolie, Tarik Dickens and Subramanian Ramakrishnan. 2017. Automated manufacturing and processing of Fiber-Reinforced Polymer (FRP) composites: An additive review of contemporary and modern techniques for advanced materials manufacturing. *Additive Manufacturing*. https://doi.org/10.1016/j.addma.2017.01.003.

Fuchs, Erica R.H., Frank R. Field, Richard Roth and Randolph E. Kirchain. 2008. Strategic materials selection in the automobile body: economic opportunities for polymer composite design. *Composites Science and Technology*. https://doi.org/10.1016/j.compscitech.2008.01.015.

Gualtieri, Luca, Ilaria Palomba, Erich J. Wehrle and Renato Vidoni. 2020. The opportunities and challenges of SME manufacturing automation: Safety and ergonomics in human-robot collaboration. In *Industry 4.0 for SMEs: Challenges, Opportunities and Requirements*. https://doi.org/10.1007/978-3-030-25425-4_4.

Gupta, Nakul et al. 2023. Advanced composite manufacturing using additive manufacturing and robotic techniques. In *E3S Web of Conferences*. https://doi.org/10.1051/e3sconf/202343001118.

Hoa, Van Suong. 2018. Automated manufacturing of composites. *Vietnam Journal of Science, Technology and Engineering*. https://doi.org/10.31276/vjste.60(1).28.

Hoffmann, Rudolf and Christoph Reich. 2023. A systematic literature review on artificial intelligence and explainable artificial intelligence for visual quality assurance in manufacturing. *Electronics (Switzerland)*. https://doi.org/10.3390/electronics12224572.

Howcroft, Debra and Phil Taylor. 2023. Automation and the future of work: A social shaping of technology approach. In *New Technology, Work and Employment*. https://doi.org/10.1111/ntwe.12240.

Hsissou, Rachid, Rajaa Seghiri, Zakaria Benzekri, Miloudi Hilali, Mohamed Rafik et al. 2021. Polymer composite materials: A comprehensive review. *Composite Structures*. https://doi.org/10.1016/j.compstruct.2021.113640.

Hu, Xueling, Alix Marcelle Sansi Seukep, Velmurugan Senthooran, Lixin Wu, Lei Wang et al. 2023. Progress of polymer-based dielectric composites prepared using fused deposition modeling 3D printing. *Nanomaterials*. https://doi.org/10.3390/nano13192711.

Huang, Dean, Derick Suarez, Putong Kang, Kornel Ehmann, Jian Cao et al. 2023. Robot forming: Automated english wheel as an avenue for flexibility and repeatability. *Manufacturing Letters*. https://doi.org/10.1016/j.mfglet.2023.08.104.

Huang, Ziqi, Yang Shen, Jiayi Li, Marcel Fey, Christian Brecher et al. 2021. A survey on AI-driven digital twins in Industry 4.0: Smart manufacturing and advanced robotics. *Sensors*. https://doi.org/10.3390/s21196340.

Hughes, Laurie, Yogesh K. Dwivedi, Nripendra P. Rana, Michael D. Williams, Vishnupriya Raghavan et al. 2022. Perspectives on the future of manufacturing within the Industry 4.0 era. *Production Planning and Control*. https://doi.org/10.1080/09537287.2020.1810762.

Hürkamp, André, Sebastian Gellrich, Tim Ossowski, Jan Beuscher, Sebastian Thiede et al. 2020. Combining simulation and machine learning as digital twin for the manufacturing of overmolded thermoplastic composites. *Journal of Manufacturing and Materials Processing*. https://doi.org/10.3390/JMMP4030092.

İpekçi, Ahmet and Bülent Ekici. 2024. An innovative composite elbow manufacturing method with 6-axis robotic additive manufacturing for fabrication of complex composite structures. *Journal of Thermoplastic Composite Materials*. https://doi.org/10.1177/08927057221109153.

Ismail, Khairul Izwan, Tze Chuen Yap and Rehan Ahmed. 2022. 3D-printed fiber-reinforced polymer composites by Fused Deposition Modelling (FDM): Fiber length and fiber implementation techniques. *Polymers*. https://doi.org/10.3390/polym14214659.

Javaid, Mohd, Abid Haleem, Ravi Pratap Singh and Rajiv Suman. 2022. Artificial Intelligence applications for Industry 4.0: A literature-based study. *Journal of Industrial Integration and Management*. https://doi.org/10.1142/S2424862221300040.

Jayasekara, Deepesh, Nai Yeen Gavin Lai, Kok Hoong Wong, Kulwant Pawar, Yingdan Zhu et al. 2022. Level of Automation (LOA) in aerospace composite manufacturing: Present status and future directions towards Industry 4.0. *Journal of Manufacturing Systems*. https://doi.org/10.1016/j.jmsy.2021.10.015.

Kappel, E., Stefaniak, D. and Hühne, C. 2013. Process distortions in prepreg manufacturing—An experimental study on CFRP L-profiles. *Composite Structures*. https://doi.org/10.1016/j.compstruct.2013.07.020.

Komarov, Dragan, Slobodan Stupar and Zoraná Posteljnik. 2011. Review of the current wind energy technologies and global market. *Journal of Applied Engineering Science*. https://doi.org/10.5937/JAES9-1120.

Kumar, Deepak, Seetharam, R. and Ponappa, K. 2024. A review on microstructures, mechanical properties and processing of high entropy alloys reinforced composite materials. *Journal of Alloys and Compounds*. https://doi.org/10.1016/j.jallcom.2023.172732.

Le, Van Nhat and Minh Tran. 2024. Recent applications of robotic manipulator in composite yacht manufacturing. In *Lecture Notes in Electrical Engineering*. https://doi.org/10.1007/978-981-99-8703-0_44.

Loeliger, Andrew, Erfu Yang and Iain Bomphray. 2021. An overview of automated manufacturing for composite materials. In *2021 26th International Conference on Automation and Computing: System Intelligence through Automation and Computing, ICAC 2021*. https://doi.org/10.23919/ICAC50006.2021.9594159.

Lowe, Brian D., Marie Hayden, James Albers and Steven Naber. 2023. Case studies of robots and automation as health/safety interventions in small manufacturing enterprises. *Human Factors and Ergonomics In Manufacturing*. https://doi.org/10.1002/hfm.20971.

Lukaszewicz, Dirk H.J.A., Carwyn Ward and Kevin D. Potter. 2012. The engineering aspects of automated prepreg layup: History, present and future. *Composites Part B: Engineering*. https://doi.org/10.1016/j.compositesb.2011.12.003.

Malaga, Miroslav, Tomas Broum, Michal Simon and Michael Fronek. 2022. Industrial robotics as an important part of modern production automation. *Acta Mechatronica*. https://doi.org/10.22306/am.v7i4.91.

Martinez-Marquez, Daniel, Nick Florin, Wayne Hall, Peter Majewski, Hao Wang et al. 2022. State-of-the-art review of product stewardship strategies for large composite wind turbine blades. *Resources, Conservation and Recycling Advances*. https://doi.org/10.1016/j.rcradv.2022.200109.

Morgan, Jeff, Mark Halton, Yuansong Qiao and John G. Breslin. 2021. Industry 4.0 smart reconfigurable manufacturing machines. *Journal of Manufacturing Systems*. https://doi.org/10.1016/j.jmsy.2021.03.001.

Mullet, Valentin, Patrick Sondi and Eric Ramat. 2021. A review of cybersecurity guidelines for manufacturing factories in Industry 4.0. *IEEE Access*. https://doi.org/10.1109/ACCESS.2021.3056650.

Naik, Nithesh, Suresh, P., Sanjay Yadav, Nisha, M.P., José Luis Arias-Gonzáles et al. 2023. A review on composite materials for energy harvesting in electric vehicles. *Energies*. https://doi.org/10.3390/en16083348.

Namsheer, K. and Chandra Sekhar Rout. 2021. Conducting polymers: A comprehensive review on recent advances in synthesis, properties and applications. *RSC Advances*. https://doi.org/10.1039/d0ra07800j.

Nejad, Amir R., Jonathan Keller, Yi Guo, Shawn Sheng, Henk Polinder et al. 2022. Wind turbine drivetrains: state-of-the-art technologies and future development trends. *Wind Energy Science*. https://doi.org/10.5194/wes-7-387-2022.

Nevliudov, Igor, Vladyslav Yevsieiev, Jalal Hasan Baker, Ayaz Ahmad, M., Vyacheslav Lyashenko et al. 2021. Development of a cyber design modeling declarative language for cyber physical production systems. *Journal of Mathematical and Computational Science*. https://doi.org/10.28919/jmcs/5152.

Parmar, Hetal, Tayyab Khan, Fausto Tucci, Rehan Umer and Pierpaolo Carlone et al. 2022. Advanced robotics and additive manufacturing of composites: Towards a new Era in Industry 4.0. *Materials and Manufacturing Processes*. https://doi.org/10.1080/10426914.2020.1866195.

Pasquale, Valentina Di, Chiara Franciosi, Raffaele Iannone and Salvatore Miranda. 2022. Special issue: Smart manufacturing for sustainability: Trends and research challenges. *Journal of Industrial Engineering and Management*. https://doi.org/10.3926/jiem.3864.

Patel, Dhruv, Chandan Kumar Sahu and Rahul Rai. 2024. Security in modern manufacturing systems: Integrating blockchain in artificial intelligence-assisted manufacturing. *International Journal of Production Research*. https://doi.org/10.1080/00207543.2023.2262050.

Peças, Paulo, Hugo Carvalho, Hafiz Salman and Marco Leite. 2018. Natural fibre composites and their applications: A review. *Journal of Composites Science*. https://doi.org/10.3390/jcs2040066.

Peng, Gang and Rahul Bhaskar. 2023. Artificial intelligence and machine learning for job automation. *Journal of Database Management*. https://doi.org/10.4018/jdm.318455.

Phanden, Rakesh Kumar, Priavrat Sharma and Anubhav Dubey. 2020. A review on simulation in digital twin for aerospace, manufacturing and robotics. In *Materials Today: Proceedings*. https://doi.org/10.1016/j.matpr.2020.06.446.

Polini, Wilma and Andrea Corrado. 2020. Digital twin of composite assembly manufacturing process. *International Journal of Production Research*. https://doi.org/10.1080/00207543.2020.1714091.

Praveena, B.A., Lokesh, N., Abdulrajak Buradi, Santhosh, N., Praveena, B.L. et al. 2022. A Comprehensive review of emerging additive manufacturing (3d printing technology): Methods, materials, applications, challenges, trends and future potential. In *Materials Today: Proceedings*. https://doi.org/10.1016/j.matpr.2021.11.059.

Priyadharshini, M., Balaji, D., Bhuvaneswari, V., Rajeshkumar, L., Sanjay, M.R. et al. 2022. Fiber reinforced composite manufacturing with the aid of artificial intelligence—A state-of-the-art review. *Archives of Computational Methods in Engineering*. https://doi.org/10.1007/s11831-022-09775-y.

Rakhshbahar, Mohammad and Michael Sinapius. 2018. A novel approach: Combination of Automated Fiber Placement (Afp) and Additive Layer Manufacturing (Alm). *Journal of Composites Science*. https://doi.org/10.3390/jcs2030042.

Rojek, Izabela, Ewa Dostatni, Dariusz Mikołajewski, Lucjan Pawłowski, Katarzyna Wegrzyn-Wolska et al. 2022. Modern approach to sustainable production in the context of Industry 4.0. *Bulletin of the Polish Academy of Sciences: Technical Sciences*. https://doi.org/10.24425/bpasts.2022.143828.

Rong, Jiang, Bi Haipeng and Zheng Ronghui. 2022. Analysis of preparation conditions of low-temperature curing powder coatings based on local clustering algorithm. *Mathematical Problems in Engineering*. https://doi.org/10.1155/2022/1143283.

Ryan, Emily A., Zach D. Seibers, John R. Reynolds and Meisha L. Shofner. 2024. Electrically conducting polymers and composites for applications in space exploration. *Journal of Applied Polymer Science*. https://doi.org/10.1002/app.55225.

Sacco, Christopher, Anis Baz Radwan, Andrew Anderson, Ramy Harik, Elizabeth Gregory et al. 2020. Machine learning in composites manufacturing: A case study of automated fiber placement inspection. *Composite Structures*. https://doi.org/10.1016/j.compstruct.2020.112514.

Sarfraz, Muhammad Salman, Hyunsoo Hong and Seong Su Kim. 2021. Recent developments in the manufacturing technologies of composite components and their cost-effectiveness in the automotive industry: A review study. *Composite Structures*. https://doi.org/10.1016/j.compstruct.2021.113864.

Sautter, Björn. 2021. Shaping digital ecosystems for sustainable production: Assessing the policy impact of the 2030 vision for Industrie 4.0. *Sustainability (Switzerland)*. https://doi.org/10.3390/su132212596.

Shackelford, Scott J. 2018. Smart factories, dumb policy? Managing cybersecurity and data privacy risks in the industrial Internet of Things. *SSRN Electronic Journal*. https://doi.org/10.2139/ssrn.3252498.

Sikka, Monica Puri, Alok Sarkar and Samridhi Garg. 2024. Artificial Intelligence (AI) in textile industry operational modernization. *Research Journal of Textile and Apparel*. https://doi.org/10.1108/RJTA-04-2021-0046.

Su, Yueyuan, Longjun Dong and Zhongwei Pei. 2022. Non-destructive testing for cavity damages in automated machines based on acoustic emission tomography. *Sensors*. https://doi.org/10.3390/s22062201.

Talib, Zainab and Abdulsamad, Jawad Hussam. 2022. Mechanical specification review of composite materials and its applications. *Journal of Current Research on Engineering*.

Villalba-Diez, Javier, Daniel Schmidt, Roman Gevers, Joaquín Ordieres-Meré, Martin Buchwitz et al. 2019. Deep learning for industrial computer vision quality control in the printing Industry 4.0. *Sensors (Switzerland)*. https://doi.org/10.3390/s19183987.

Wang, Weiyu and Keng Siau. 2019. Artificial intelligence, machine learning, automation, robotics, future of work and future of humanity: A review and research Agenda. *Journal of Database Management*. https://doi.org/10.4018/JDM.2019010104.

Wang, Xin, Man Jiang, Zuowan Zhou, Jihua Gou, David Hui et al. 2017. 3D Printing of polymer matrix composites: A review and prospective. *Composites Part B: Engineering*. https://doi.org/10.1016/j.compositesb.2016.11.034.

Wang, Yuan Bin, Pai Zheng, Tao Peng, Hua Yong Yang, Jun Zou et al. Smart additive manufacturing Current artificial intelligence-enabled methods and future perspectives. *Science China Technological Sciences*. https://doi.org/10.1007/s11431-020-1581-2.

Wiśnicki, Bogusz, Natalia Wagner and Piotr Wołejsza. 2024. Critical areas for successful adoption of technological innovations in sea shipping—The autonomous ship case study. *Innovation: The European Journal of Social Science Research*. https://doi.org/10.1080/13511610.2021.1937071.

Witik, Robert A., Fabrice Gaille, Rémy Teuscher, Heike Ringwald, Véronique Michaud et al. 2012. Economic and environmental assessment of alternative production methods for composite aircraft components. *Journal of Cleaner Production*. https://doi.org/10.1016/j.jclepro.2012.02.028.

Wood, Nathan D., Lisa J. Gillie, David J. Cooke and Marco Molinari. 2022. A review of key properties of thermoelectric composites of polymers and inorganic materials. *Materials*. https://doi.org/10.3390/ma15238672.

Woschank, Manuel, Erwin Rauch and Helmut Zsifkovits. 2020. A review of further directions for artificial intelligence, machine learning, and deep learning in smart logistics. *Sustainability (Switzerland)*. https://doi.org/10.3390/su12093760.

Yassin, Khaled and Mehdi Hojjati. 2018. Processing of thermoplastic matrix composites through automated fiber placement and tape laying methods: A review. *Journal of Thermoplastic Composite Materials*. https://doi.org/10.1177/0892705717738305.

Zhang, Mingyue, Gill M. Biesold, Woosung Choi, Jiwoo Yu, Yulin Deng, Clara Silvestre and Zhiqun Lin. 2022. Recent advances in polymers and polymer composites for food packaging. *Materials Today*. https://doi.org/10.1016/j.mattod.2022.01.022.

Zhang, Yunbo and Tsz Ho Kwok. 2018. Design and interaction interface using augmented reality for smart manufacturing. In. https://doi.org/10.1016/j.promfg.2018.07.140.

Zheng, Pai, Honghui Wang, Zhiqian Sang, Ray Y. Zhong, Yongkui Liu et al. 2018. Smart manufacturing systems for Industry 4.0: Conceptual framework, scenarios, and future perspectives. *Frontiers of Mechanical Engineering*. https://doi.org/10.1007/s11465-018-0499-5.

Zhong, Ray Y., Xun Xu, Eberhard Klotz and Stephen T. Newman. 2017. Intelligent manufacturing in the context of Industry 4.0: A review. *Engineering*. https://doi.org/10.1016/J.ENG.2017.05.015.

CHAPTER

10 Material and Manufacturing Aspect of Flexible Robots

Nishu Sharma,[1] *Nathi Ram Chauhan*[2,*]
and *Akshay Kumar Shakya*[3]

The advancement of flexible robots has transformed sectors like healthcare, manufacturing, and exploration by enhancing adaptability, safety, and functional versatility. A key enabler is the strategic use of materials that offer flexibility, durability, and responsiveness. This chapter reviews recent innovations in materials such as soft polymers, silicone rubber, polyurethanes, hydrogels, shape-memory polymers/alloys, IPMCs, and bio-inspired composites. It also examines mechanical performance, fabrication strategies, and application-specific material suitability. Emerging trends like self-healing and sensing through smart materials are highlighted, along with challenges tied to scalability, biocompatibility, and environmental impact. The chapter concludes by identifying research gaps and proposing future directions to optimize materials for next-generation flexible robotics.

[1] Research Scholar, Department of Mechanical & Automation Engineering, Indira Gandhi Delhi Technical University for Women, Delhi and Assistant Professor, Mechanical Engineering, Mangalmay Institue of Management & Technology, Greater Noida, India.

[2] Professor, Department of Mechanical & Automation Engineering, Indira Gandhi Delhi Technical University for Women, Delhi, India.

[3] Assistant Professor, Mechanical Engineering, Sanskar College of Engineering & Technology, Ghaziabad, India.

* Corresponding author: nrchauhan@igdtuw.ac.in

DOI: 10.1201/9781003607359-10

10.1 Introduction

Flexible robots are transforming industries by offering adaptability, precision, and efficiency in environments requiring compliance and dynamic interaction. Unlike rigid systems, soft robots safely interact with their surroundings, conform to complex geometries, and perform tasks in constrained or unpredictable conditions. Their bio-inspired design makes them essential across healthcare, agriculture, and advanced manufacturing systems. Figure 10.1 illustrates industrial applications of soft robots—(a) a spherical-type gripper and (b) a four-finger gripper—while Fig. 10.2 shows their use in medical rehabilitation, featuring (a) a surgical robot and (b) robotic gloves (Zhao et al., 2021).

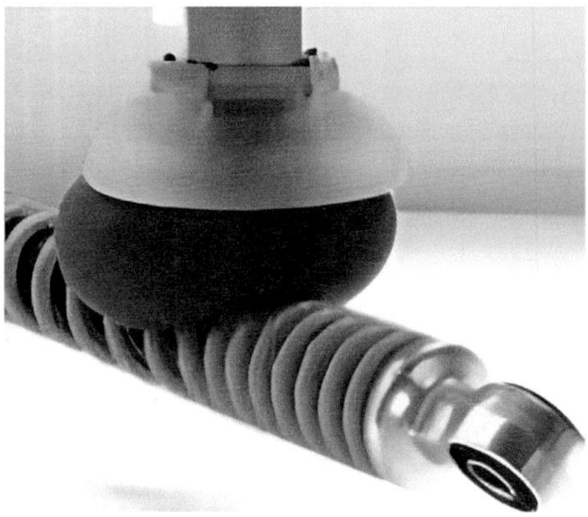

Figure 10.1(a): A spherical-type soft gripper (Zhao et al., 2021).

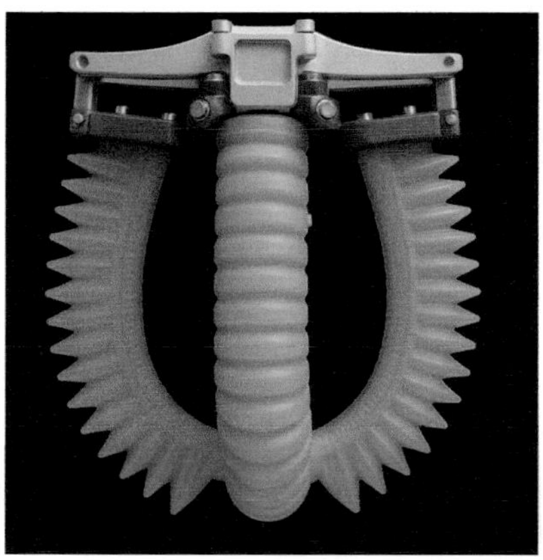

Figure 10.1(b): A four-finger-type soft gripper (Zhao et al., 2021).

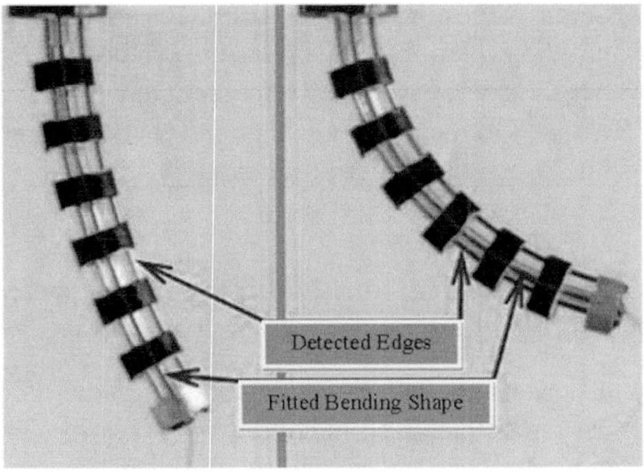

Figure 10.2(a): A surgical robot used in medical (Zhao et al., 2021).

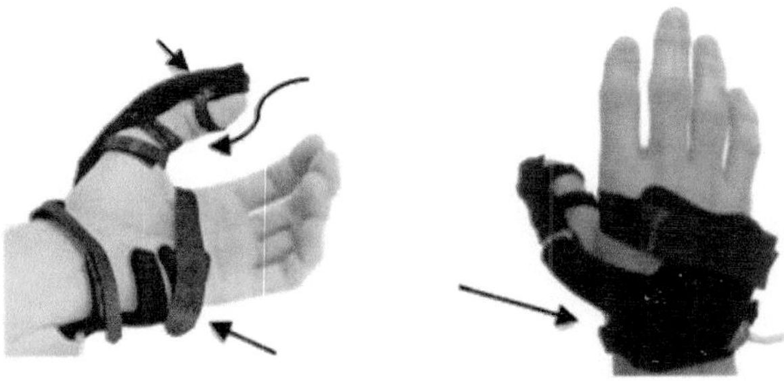

Figure 10.2(b): Robotic gloves used in medical (Zhao et al., 2021).

Flexible robots play a crucial role in Flexible manufacturing systems (FMS), which emphasize adaptability and customization in production workflows. These robots enhance precision and agility for tasks like assembly, sorting, and material handling (Groover, 2019). Integrated into cooperative robotic systems, they enable multi-robot collaboration, allowing diverse units to work together to optimize production efficiency (Nof, 1999).

The effectiveness of flexible robots largely depends on two key elements: material selection and manufacturing techniques. Components like actuators, sensors, and structural parts require materials with high elasticity, durability, and often responsiveness to stimuli. Their fabrication involves both conventional and advanced methods—such as additive manufacturing and multi-material molding—to create complex, integrated designs (Wang et al., 2021; Giberti et al., 2022). Manufacturing flexible robots presents key challenges. Material selection must balance strength, flexibility, and environmental stability. Ensuring repeatability, precision, and functional integration demands scalable, cost-effective techniques. Overcoming these hurdles is crucial to advancing flexible robotic systems.

This chapter examines the material and manufacturing foundations of flexible robots, emphasizing their pivotal role in enabling cooperative and adaptive manufacturing systems. Drawing from key references like Flexible Manufacturing of Cooperating Robots (Groover, 2019), it underscores how advancements in manufacturing technologies and material innovation drive robotic functionality, integration, and performance.

10.1.1 Importance of Material and Manufacturing Choices in Flexible Robotic Systems

The versatility and performance of flexible robotic systems largely depend on material selection and manufacturing methods. These choices affect the robot's adaptability, environmental interaction, and reliability under varying conditions. In highly dynamic and customizable production settings, soft robotics plays a vital role in achieving manufacturing flexibility (Slack, 1987).

10.1.2 Role of Materials in Flexibility

Materials form the backbone of flexible robotic systems, enabling adaptive, efficient motion. Key properties include elasticity, resilience, low weight, and durability to endure repeated deformation without losing integrity. For healthcare and soft robotics, biocompatible materials like hydrogels and silicones are essential. Smart materials such as electroactive polymers and shape-memory alloys offer responsiveness for sensing and actuation. Tailoring these properties to specific tasks allows engineers to optimize robotic performance in collaborative and precision-driven operations.

10.1.3 Manufacturing's Impact on Flexibility

Manufacturing flexible robotic systems is complex due to multifunctional components and specialized material requirements. Advanced techniques like 3D printing, molding, and additive manufacturing enable rapid prototyping and customization. These methods must meet specific operational demands while integrating multiple functions—such as actuation and sensing—into single structures. Scalability remains critical, supporting production from micro- to large-scale systems. Compatibility between material properties and fabrication techniques directly impacts robotic flexibility. For precision and uniformity, elastomers are commonly used in molding flexible actuators.

10.1.4 Flexibility as a Manufacturing Objective

Flexible robotic systems embody manufacturing flexibility by enabling rapid adaptation to shifting product designs and customer demands. As Slack (1987) emphasizes, flexibility is essential for maintaining competitiveness in evolving markets. Through advanced materials and modern fabrication methods, flexible robots enhance production efficiency, responsiveness, and reduce lead times.

The synergy between manufacturing processes and material selection is key to designing effective flexible robotic systems. These choices impact not only cost, scalability, and reliability but also define functionality and mechanical performance—crucial for achieving manufacturing flexibility in competitive industrial settings.

10.2 Material Requirements for Flexible Robots

Soft robots exhibit flexibility, durability, and functional adaptability, enabling them to operate in dynamic and complex environments. These properties depend heavily on material selection. As noted by Kopacek et al. (1988) and Wang et al. (2021), material characteristics directly influence mechanical adaptability and long-term performance. Elasticity supports stretching, bending, and load-bearing without permanent deformation, while fatigue resistance ensures durability under repeated stress. For biomedical and wearable applications, materials like hydrogels and silicones offer biocompatibility for safe human interaction. In contrast, industrial and exploration uses demand thermal and chemical resistance to ensure stability in extreme conditions.

10.2.1 Material Categories

Soft robotics focuses on deformability and flexibility, mimicking biological organisms through the use of materials such as polymers, hydrogels, elastomers, composites, smart materials, and shape-memory alloys. These materials are categorized by their mechanical and functional properties to meet diverse application needs.

10.2.1.1 Elastomers

Elastomers, including materials like silicone rubber and polyurethanes, are key to flexible robotics. Silicone rubber stands out for its flexibility, thermal stability, and biocompatibility—making it ideal for wearable devices, soft grippers, and artificial muscles. Recent advancements in silicone composites enhance electrical conductivity and mechanical strength, enabling seamless integration into components such as actuators and sensors.

Polyurethanes are durable, highly elastic elastomers with strong abrasion resistance, making them ideal for stretchable electronics, medical devices, and soft actuators. Current research focuses on improving their biocompatibility and developing eco-friendly alternatives to minimize environmental impact without compromising performance.

10.2.1.2 Gels

Gels, including hydrogels and Ionic Polymer-Metal Composites (IPMCs), are widely used in soft robotics due to their flexibility and functional properties. Hydrogels are especially valued for their high water content and biocompatibility, enabling large deformations—ideal for applications like drug delivery, tissue engineering, and bio-inspired robots. Recent research focuses on tough, self-healing hydrogels responsive to stimuli such as light, pH, and temperature. These next-generation hydrogels are now employed in biomedical robots, soft actuators, and artificial muscles (Wang et al., 2021). Ionic Polymer-Metal Composites (IPMCs) are valued for their lightweight, flexibility, and electromechanical coupling, making them ideal for actuators, soft sensors, and artificial muscles. Recent advancements have improved their durability and actuation performance, with ongoing research developing novel metal electrodes for enhanced longevity and efficiency. These innovations support their expanding use in biomedical and soft robotics applications.

10.2.1.3 Shape-Memory Alloys (SMAs) and Polymers (SMPs) or Smart Materials

Shape Memory Alloys (SMAs) are smart materials known for their elasticity—returning to their original shape after deformation when exposed to heat or electric fields. Wang et al. (2021) highlight their high strength and stimuli-responsive behavior, making them ideal for soft actuators, deployable structures, and medical devices. Recent advances focus on SMAs with lower activation temperatures and faster response times, expanding their efficiency and applications in biomedical and technological domains. Shape-Memory Polymers (SMPs) are lightweight, highly deformable materials that return to their original shape when heated. They are used in morphing structures, soft grippers, and self-healing systems. Recent advancements have improved recovery speed and introduced multi-responsive SMPs capable of reacting to multiple stimuli simultaneously. These innovations expand their use in biomedical devices, smart materials, and advanced robotics.

10.2.1.4 Composite Materials

Composite materials combine two or more substances with distinct physical and chemical properties to achieve enhanced functionality. Examples include elastomer-metal composites, which merge flexibility with electrical conductivity, making them suitable for soft robotics, wearable electronics, and stretchable sensors. Recent research focuses on improving mechanical strength and developing multifunctional composites that integrate actuators, sensors, and energy storage components. Nano-composites combine two or more materials at the nanoscale, offering superior mechanical strength, and enhanced thermal and electrical conductivity. These properties make them ideal for energy storage, sensors, and soft actuators. Types include graphene-based, ceramic matrix, polymer, and carbon nanotube composites. Incorporating nanomaterials like graphene and CNTs significantly boosts efficiency, responsiveness, and durability in fields such as bio-inspired robotics, advanced energy systems, and flexible electronics.

10.2.1.5 Liquid Crystal Elastomers (LCEs)

Liquid Crystal Elastomers (LCEs) are smart soft materials that change shape in response to stimuli like light and temperature. Their unique responsiveness makes them ideal for applications such as soft actuators, artificial muscles, and adaptive surfaces. Recent research focuses on enhancing their efficiency, flexibility, and responsiveness, broadening their use in biomedical devices, soft robotics, and smart materials.

10.2.1.6 Biodegradable and Biocompatible Materials

Biodegradable and biocompatible materials include natural and synthetic polymers. Natural polymers offer biodegradability, low toxicity, and compatibility with biological tissues, making them ideal for bio-inspired robotics, medical devices, and tissue engineering. Recent research focuses on enhancing durability and mechanical strength through composite development, expanding their role in soft robotics and biomedical applications.

Synthetic biodegradable polymers—such as polyesters, polyphosphazenes, polyanhydrides, and polyurethanes—are widely used in tissue engineering due to their tunable mechanical properties and controlled degradation rates. These features make them suitable for temporary implants, medical devices, and environmental sensors. Recent advances have improved biocompatibility and enabled precise control over degradation, supporting their growing use in sustainable technologies and eco-friendly medical applications.

10.2.1.7 Electroactive Polymers (EAPs)

Electroactive Polymers (EAPs) are conductive materials that change shape or size in response to electric fields. Key types include dielectric elastomers and conductive polymers. Dielectric elastomers, known for their lightweight and smart behavior, deform under electrical stimulation and are widely used in adaptive optics, soft actuators, and artificial muscles. Recent research focuses on improving energy efficiency, lowering actuation voltages, increasing dielectric constants, and expanding their use in next-generation smart materials and soft robotics.

Conductive polymers are organic materials that offer both electrical conductivity and mechanical flexibility, making them ideal for soft sensors, actuators, and flexible electronics. Recent advancements focus on enhancing their strength and conductivity through integration with nanocomposites like graphene and carbon nanotubes. These improvements have broadened their applications in soft robotics, wearable tech, and bio-integrated systems.

10.2.2 Selection Criteria

Material selection is critical for soft robotics, as it directly affects performance, safety, and durability. Key factors include elasticity, flexibility, light weight, fatigue resistance, and environmental compatibility. As noted by Kopacek et al. (1988), soft actuators require materials with both elasticity and load-bearing capacity to preserve structural integrity. Biocompatibility and sterilizability are essential for biomedical use, while lightweight yet strong materials are vital for wearable and mobile robots (Wang et al., 2021). Selecting materials aligned with specific operational demands enhances both functionality and efficiency in flexible robotic systems.

10.3 Manufacturing Techniques for Flexible Robots

Manufacturing techniques for soft robots have have evolved to meet the unique demands of adaptive and flexible systems. These methods enable the integration of complex geometries, sensors, and actuators within flexible structures. Drawing on insights from Giberti et al. (2021) and Wang et al. (2020), this section outlines the key fabrication approaches tailored to soft and flexible robotics.

10.3.1 Additive Manufacturing (3D Printing)

Additive manufacturing (AM), or 3D printing, is a bottom-up fabrication method that builds objects layer by layer—ideal for producing complex, customized flexible robotic components that mimic biological structures. Unlike traditional techniques, AM enables intricate geometries and rapid prototyping, streamlining design iteration and minimizing material waste (Giberti et al., 2021). Materials like Thermoplastic Elastomers (TPE) and Thermoplastic Polyurethane (TPU) are commonly used due to their elasticity and durability. Key AM techniques include Stereolithography (SLA), Material Jetting (MJ), Fused Deposition Modelling (FDM), Powder Bed Fusion (PBF), and Direct Ink Writing, each suited for specific soft robotic applications (Chan et al., 2012).

10.3.1.1 Fused Deposition Modelling (FDM)

Fused Filament Fabrication (FFF), also known as Fused Deposition Modelling (FDM), is a widely used 3D printing technique that extrudes molten thermoplastics—such as ABS, PLA, nylon, and PC—to fabricate solid structures. It is especially favored for producing soft robotic components and actuators due to its cost-effectiveness and design flexibility. Anver et al. (2017) demonstrated FDM's utility in printing soft fingers and grippers using FilaFlex and Ninjaflex, while Mutlu et al. (2016) integrated stiffness-enhancing units and electroactive composites. Notable innovations include bend-twist actuators, shape memory polymer (SMP) grippers, and industrial soft grippers. Figure 10.3.2 illustrates

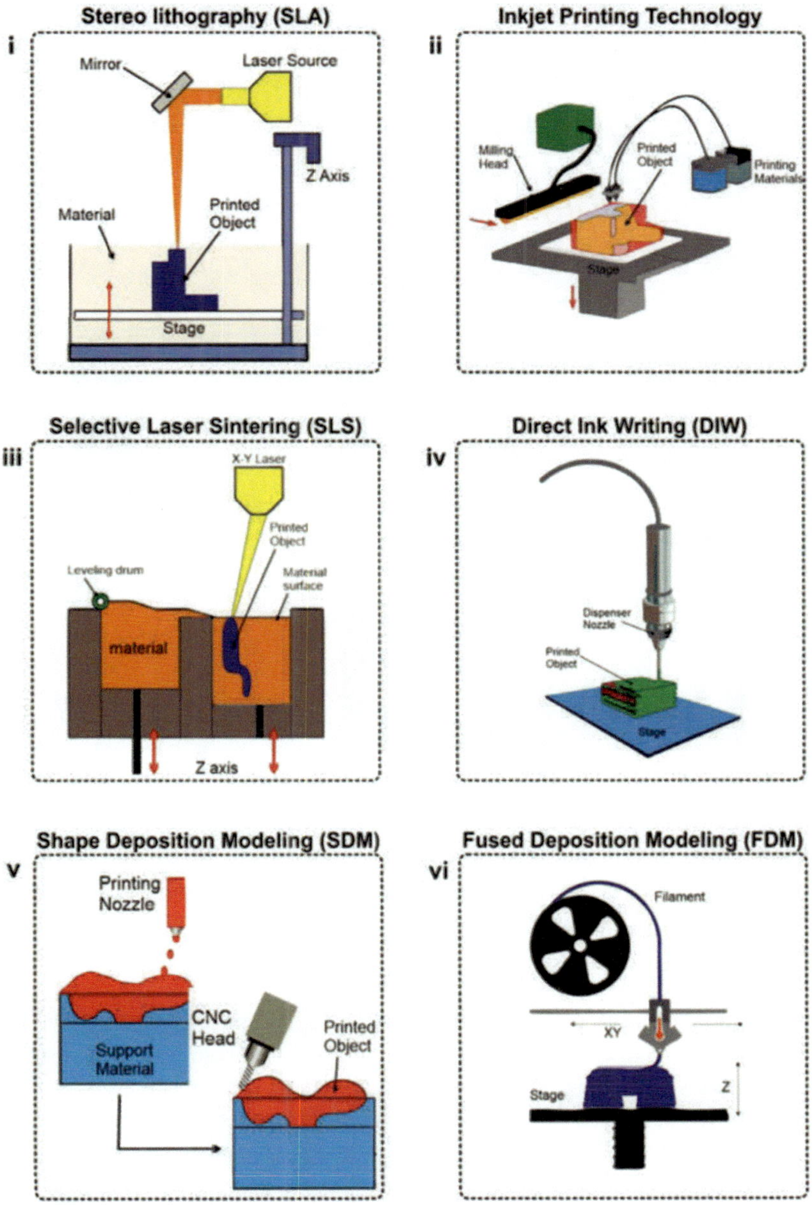

Figure 10.3.1: Different techniques of 3D printing (Jahan zeb gul, 2018).

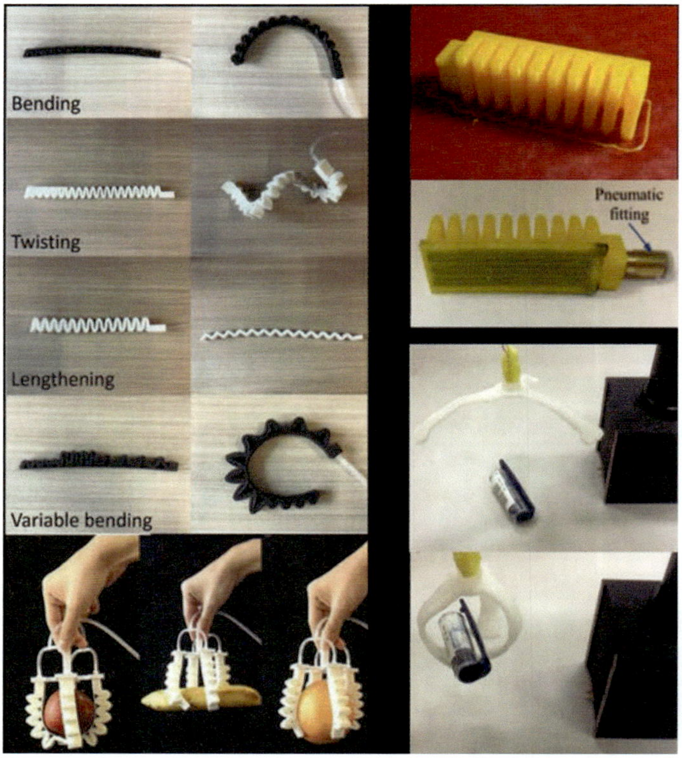

Figure 10.3.2: Soft robots printed by FDM (Keong and Hua, 2018; Elgeneidy et al., 2018; Yang et al., 2016).

examples from Keong and Hua (2018), Elgeneidy et al. (2018), and Yang et al. (2016), showcasing design versatility and functional integration in FDM-based soft robotics.

10.3.1.2 Vat Photopolymerization

Vat Photopolymerization—encompassing DLP, 2PP, and SLA—uses UV light or lasers to cure photopolymer resins, enabling high-resolution 3D printing. These methods are widely applied in fabricating soft robots, actuators, and bio-integrated devices, such as SLA-printed hydrogel robots actuated by cardiac cells and DLP-printed actuators using commercial resins (Chan et al., 2012). Advanced work includes electroactive hydrogels for locomotion and drug delivery via magnetic-field-assisted stereolithography (Joyee and Pan, 2019). Figure 10.3.3 showcases stereolithography-based soft robots: (a) flexible grippers with TangoPlus (Ge et al., 2018), (b) Spot-E printed fingers (Bryan et al., 2015), and (c) EAH-based grippers actuated electrically (Han et al., 2018a).

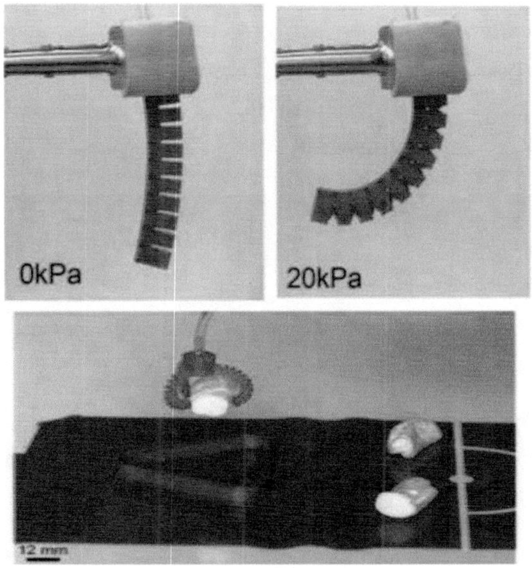

Figure 10.3.3(a): A flexible finger and gripper printed with TangoPlus from Stratasys (Ge et al., 2018).

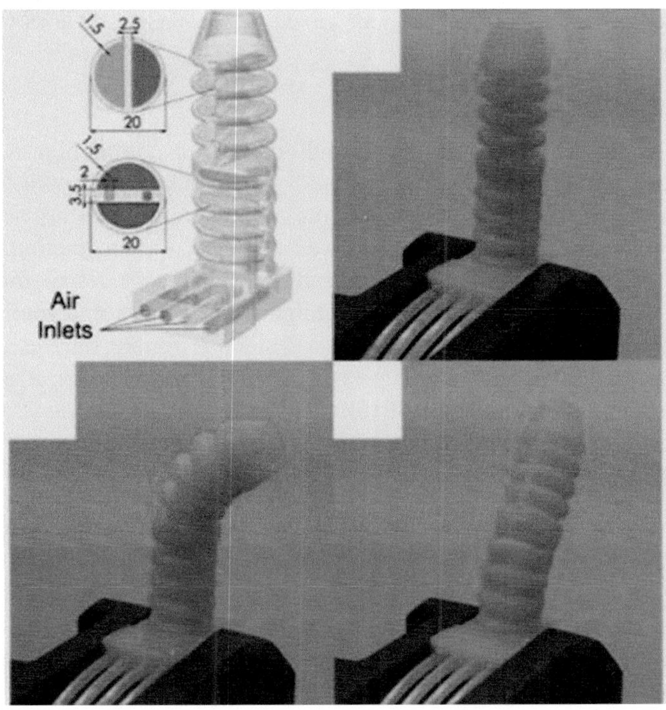

Figure 10.3.3(b): Flexible fingers printed using Spot-E (Bryan et al., 2015).

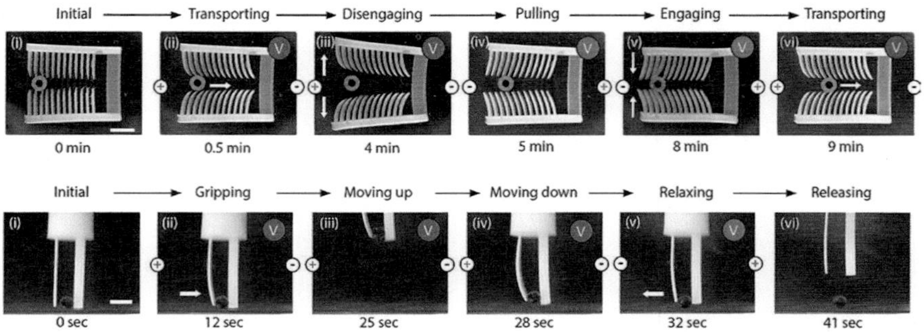

Figure 10.3.3(c): A gripper and transporter fabricated with electroactive hydrogels (Han et al., 2018a).

10.3.1.3 Powder Bed Fusion – Selective Laser Sintering (SLS)

Selective Laser Sintering (SLS) uses electron beams or lasers to fuse polymers like PA 12 and TPU, enabling highly flexible and customizable 3D-printed designs. In soft robotics, it supports the fabrication of adaptive grippers, robotic hands, and bioinspired actuators—such as FESTO's Bionic Handling Assistant, which emulates an elephant's trunk. Roppenecker et al. (2013) and Jäntsch et al. (2013) showcased SLS-fabricated multi-fingered hands, snake-like manipulators, and anthropomimetic arms with optimized tendon pathways. SLS's capacity for intricate geometries and reusable supports makes it pivotal for soft and bioinspired robotic innovations.

10.3.1.4 Material Jetting

Material jetting, including polyjet and multijet methods, offers high-resolution, multi-material 3D printing by jetting liquid resin and curing it with UV light. It enables the fabrication of soft robots with gradient stiffness—such as combustion-powered designs and bellowed actuators for locomotion (Bartlett et al., 2015). Zhang et al. (2019) utilized shape memory resins for tunable actuators, while Rossiter et al. (2009) highlighted challenges in compliance and fatigue. Recent advances like drop-on-demand silicone printing have expanded its use in rehabilitation robotics, making it a powerful tool for complex, adaptive soft robotic systems (Jahan Zeb et al., 2016).

10.3.2 Moulding and Casting

Moulding and casting remain key techniques for fabricating flexible robotic components with precision. Silicone casting is widely used for actuators due to its elasticity and biocompatibility, enabling fine control over geometry and material thickness—essential for flexibility and performance (Wang et al., 2020). Multi-material moulding enhances integration by combining rigid and flexible parts in a single mould, reducing assembly time and improving function. Similarly, layered

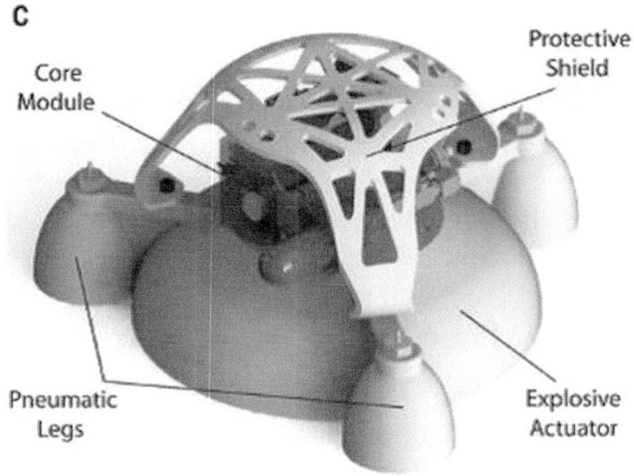

Figure 10.3.4(a): A gradient stiffness soft robot powered by combustion (Bartlett et al., 2015).

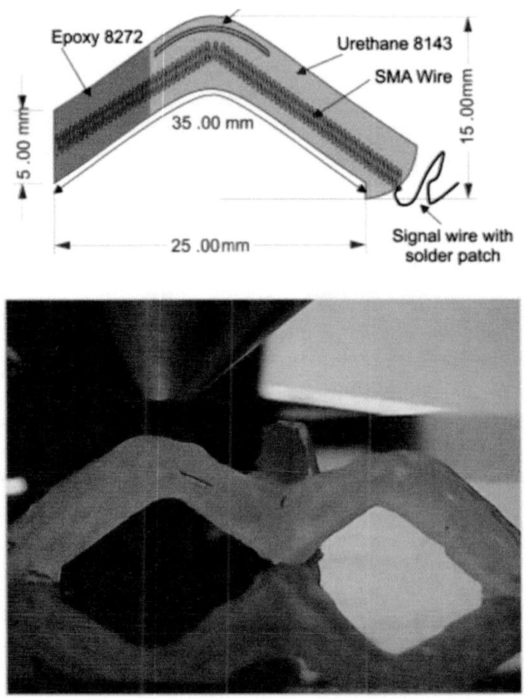

Figure 10.3.4(b): A tri-pedal soft robot printed using two flexible materials, epoxy and polyurethane (Jahan Zeb et al., 2016).

and lamination fabrication—used for soft skins, circuits, and sensors—stacks thin materials to create multifunctional, adaptive structures. This supports stretchable sensors, wearable robotics, and flexible electronics across diverse applications.

10.3.3 Hybrid Manufacturing Approaches

Hybrid manufacturing combines additive and subtractive techniques to produce complex robotic components with diverse material properties. It enables precise fabrication, such as 3D-printed molds for tailored soft actuators and integration of CNC-machined rigid parts with flexible elements. As noted by Giberti et al. (2021), this approach is both cost-effective and performance-driven, offering optimized solutions for flexible robotics.

10.4 Challenges in Material and Manufacturing Aspects

Flexible robotic systems face key challenges in material selection and fabrication, driven by the need for high performance, cost-effectiveness, scalability, and sustainability. Polymers and elastomers, though versatile, degrade over time due to mechanical fatigue, environmental exposure, and chemical wear, reducing their functionality. Micro-cracks from repeated stress and accelerated aging from UV and temperature fluctuations further limit durability. Solutions such as surface coatings and improved formulations are being developed to enhance longevity (Hasanov et al., 2021; Kumar and Sathiya, 2020). Maintaining uniformity in flexible parts is also difficult—multi-material additive manufacturing can lead to weak bonding and dimensional inconsistencies. Addressing these issues requires advanced quality control and optimized processing techniques (Kanishka and Acherjee, 2021).

Scaling up flexible robot production cost-effectively remains a major challenge. Techniques like silicone molding and multi-material 3D printing demand specialized equipment, raising costs (Mäntylä et al., 1996). Additionally, high costs of smart materials—like electroactive polymers and shape memory alloys—limit widespread adoption (Kumar and Sathiya, 2020). As noted by Kanishka and Acherjee (2021), embedding sensors and actuators into soft structures can create rigid zones, reducing flexibility. To address this, researchers are developing stretchable conductors and soft circuits that integrate seamlessly without compromising performance (Hasanov et al., 2021). Recyclability and sustainability are growing priorities in flexible robotics. Many polymers used are derived from non-renewable sources and present recycling challenges, while additive manufacturing generates significant material waste (Kumar and Sathiya, 2020). To reduce environmental impact, researchers are pursuing biodegradable materials and closed-loop recycling systems. However, no single material fulfills all demands for flexibility, strength, and durability. Conventional simulation-based models increase computational costs and prolong development, while traditional sensors struggle with tracking internal deformation, limiting real-time control. Despite 3D printing advances, material limitations persist. Additionally, current energy sources remain bulky, constraining miniaturization. Balancing flexibility and rigidity, especially for micro-robotics, continues to be a major challenge (Zhao, 2021).

10.5 Recent Advances and Innovations in Material and Manufacturing for Flexible Robots

Recent advances in flexible robotics are powered by bio-inspired materials, smart materials, and multi-material 3D printing. Muscle-like elastomers and hydrogels enhance adaptability and motion efficiency (Rostamabadi et al., 2021), while synthetic skin materials with self-healing and tactile sensing improve human-robot interaction. Smart materials such as electroactive polymers and shape memory alloys support actuation via thermal and electrical input (Tan et al., 2021). Self-adaptive materials responsive to moisture, heat, or light further boost real-world functionality. Multi-material 3D printing enables seamless integration of flexible and rigid parts, advancing the development of soft grippers, exoskeletons, and sensors with superior performance (Hu et al., 2021). Artificial intelligence (AI) and machine learning (ML) play a pivotal role in advancing material discovery and robotic design. AI-driven algorithms accelerate development by predicting novel material compositions with desired traits like biocompatibility, durability, and elasticity (Shaplov et al., 2021). ML techniques optimize structural designs by analyzing large datasets to balance strength and flexibility. Additionally, AI-powered predictive maintenance models assess material degradation and recommend preventive measures, enhancing robotic longevity. These innovations are bridging advanced engineering with bio-inspired adaptability in flexible robotic systems.

10.6 Case Studies: Successful Flexible Robots in Various Industries

Soft robotics has transformed multiple industries by overcoming the limitations of rigid robots. In engineering, soft grippers have enhanced automation in sectors needing delicate handling—such as Ocado's warehouses, where they safely manage fragile items like vegetables and electronics. These grippers, made from adaptable materials like silicone and hydrogels, use pneumatic actuation to adjust grip force, ensuring gentle and damage-free handling (Majidi, 2021). Their success in food and electronics sectors underscores their role in reducing waste and boosting efficiency.

In healthcare, wearable robots and exoskeletons have transformed mobility and rehabilitation for individuals with physical impairments. Ekso Bionics has developed bio-inspired exoskeletons that aid recovery from spinal cord injuries and neurological conditions, offering comfort and natural movement (Dheerthi et al., 2021). Flexible electronics integrated into these systems enable real-time monitoring of joint angles and muscle activity, enhancing personalized physical therapy (Heng et al., 2021). These innovations improve rehabilitation efficiency and patient outcomes. Flexible robotics has shown remarkable adaptability in harsh environments. NASA has developed soft robotic prototypes using bio-inspired materials like shape memory polymers to navigate icy and rugged terrains in space (Majidi, 2021). These robots, equipped with soft appendages, are also ideal for maneuvering in confined or irregular spaces—supporting disaster response and deep-sea exploration (Zhu et al., 2021). Their integration into environmental and space missions enhances scientific research and recovery operations beyond traditional robotic capabilities.

10.7 Future Directions in Material and Manufacturing for Flexible Robots

The future of soft robotics will be shaped by progress in materials, computation, scalable manufacturing, and sustainability. Advanced materials like shape-memory alloys, electroactive polymers, and bio-inspired polymers will enhance actuation, resilience, and adaptability (Zhu et al., 2021; Tan et al., 2021). Additionally, new soft composites with high strength-to-weight ratios will enable robots to perform complex, heavy-duty tasks while remaining flexible in dynamic environments. These innovations will be key to expanding the capabilities of flexible robotic systems. Scaling up flexible robot manufacturing for commercial use remains a key challenge. Combining 3D printing with traditional methods like injection moulding offers a cost-effective way to produce complex parts (Heng et al., 2021). AI-driven automation further enhances scalability and reduces costs (Majidi, 2021). Emphasis on sustainable robotics—including recyclability, energy-efficient actuation, and biodegradable materials—aims to lower environmental impact (Kumar and Sathiya, 2020). With AI and machine learning optimizing material selection and design, next-generation soft robots will be smarter, more efficient, and eco-conscious, shaping future industrial automation.

10.8 Conclusion

Soft robotics represents a breakthrough in material science and manufacturing. Built from flexible materials like polymers, hydrogels, elastomers, composites, smart materials, and shape memory alloys, soft robots offer biocompatibility, elasticity, and durability—allowing them to operate effectively in dynamic environments. Soft robotic components are manufactured using advanced techniques like 3D printing, moulding, and hybrid fabrication. Despite these innovations, challenges persist—such as scalability, material degradation, and sustainability. Overcoming these issues with durable materials and cost-effective methods is key to broader adoption in industry, healthcare, and exploration. Already, flexible robots are proving valuable in applications like soft grippers, rehabilitation wearables, and adaptive systems for extreme environments.

Future research in soft robotics will focus on AI-driven design, smart actuators, and bio-inspired materials to develop eco-friendly, efficient, and commercially viable systems. Seamless integration of manufacturing techniques and material properties will be key to balancing flexibility and structural strength. Combining rigid and soft materials in a single design, supported by advanced computational tools for pre-prototyping optimization, will enhance both performance and efficiency. Ultimately, a multidisciplinary approach—merging intelligent design, materials science, and scalable manufacturing—will shape the next generation of adaptive, sustainable soft robots.

References

Anver, Hisham M.C.M., Rahim Mutlu and GurselAlici. 2017. 3D printing of a thin-wall soft and monolithic gripper using fused filament fabrication. In *2017 IEEE International Conference on Advanced Intelligent Mechatronics (AIM)*, pp. 442–447. IEEE.

Appiah, C., Arndt, C., Siemsen, K., Heitmann, A., Staubitz, A. et al. 2019. Living materials herald a new era in soft robotics. *Advanced Materials*, 31(36): 1807747.

Bartlett, N.W., Tolley, M.T., Overvelde, J.T.B., Weaver, J.C., Mosadegh, B., Bertoldi, K., Whitesides, G.M. and Wood, R.J. 2015. A 3D-printed, functionally graded soft robot powered by combustion. *Science*, 349(6244): 161.

Case, J.C., White, E.L. and Kramer, R.K. 2015. Soft material characterization for robotic applications. *Soft Robotics*, 2(2): 80–87.

Chan, Vincent, Jae Hyun Jeong, Piyush Bajaj, Mitchell Collens, TaherSaif, Hyunjoon Kong et al. 2012. Multi-material bio-fabrication of hydrogel cantilevers and actuators with stereolithography. *Lab on a Chip*, 12(1): 88–98.

Chossat, Jean-Baptiste, Yong-Lae Park, Robert J. Wood and Vincent Duchaine. 2013. A soft strain sensor based on ionic and metal liquids. *IEEE Sensors Journal*, 13(9): 3405–3414.

Coyle, S., Majidi, C., LeDuc, P. and Hsia, K.J. 2018. Bio-inspired soft robotics: Material selection, actuation, and design. *Extreme Mechanics Letters*, 22: 51–59.

Dämmer, Gabriel, Sven Gablenz, Alexander Hildebrandt and Zoltan Major. 2019. PolyJet-printed bellows actuators: Design, structural optimization, and experimental investigation. *Frontiers in Robotics and AI*, 6: 34.

Dheerthi, N., Kumar, A.K., Sarveswaran, S. and Murugarajan, A. 2024. A comprehensive analysis of bio-inspired soft robotic ARM for healthcare applications. In *Multidisciplinary Applications of AI Robotics and Autonomous Systems* (pp. 1–23). IGI Global.

Elango, N. and Faudzi, A.A.M. 2015. A review article: Investigations on soft materials for soft robot manipulations. *The International Journal of Advanced Manufacturing Technology*, 80: 1027–1037.

Giberti, H., Abbattista, T., Carnevale, M., Giagu, L., Cristini, F. et al. 2022. A methodology for flexible implementation of collaborative robots in smart manufacturing systems. *Robotics*, 11(1): 9.

Groover, M.P. 2016. *Automation, Production Systems, and Computer-integrated Manufacturing*. Pearson Education India.

Gul, Jahan Zeb, MemoonSajid, Muhammad MuqeetRehman, Ghayas Uddin Siddiqui, Imran Shah et al. 2018. 3D printing for soft robotics—A review. *Science and Technology of Advanced Materials*, 19(1): 243–262.

Hasanov, S., Alkunte, S., Rajeshirke, M., Gupta, A., Huseynov, O. et al. 2021. Review on additive manufacturing of multi-material parts: Progress and challenges. *Journal of Manufacturing and Materials Processing*, 6(1): 4.

Heng, W., Solomon, S. and Gao, W. 2022. Flexible electronics and devices as human–machine interfaces for medical robotics. *Advanced Materials*, 34(16): 2107902.

Hu, T., Wang, Z., Shen, W., Liang, R., Yan, D. et al. 2021. Recent advances in innovative strategies for enhanced cancer photodynamic therapy. *Theranostics*, 11(7): 3278.

Iida, F. and Laschi, C. 2011. Soft robotics: Challenges and perspectives. *Procedia Computer Science*, 7: 99–102.

Jahan Zeb, G., Bong-Su, Y., Young Jin, Y., Dong Eui, C. and Kyung Hyun, C. 2016. *In situ* UV curable 3D printing of multi-material tri-legged soft bot with spider mimicked multi-step forward dynamic gait. *Smart Materials and Structures*, 25: 115009.

Jiao, Z., Zhang, C., Wang, W., Pan, M., Yang, H. et al. 2019. Advanced artificial muscle for flexible material-based reconfigurable soft robots. *Advanced Science*, 6(21): 1901371.

Kanishka, K. and Acherjee, B. 2023. Revolutionizing manufacturing: A comprehensive overview of additive manufacturing processes, materials, developments, and challenges. *Journal of Manufacturing Processes*, 107: 574–619.

Kopacek, P., Desoyer, K. and Lugner, P. 1989. Modelling of flexible robots—An introduction. In *Robot Control* 1988 (Syroco'88) (pp. 21–28). Pergamon.

Kumar, M.B. and Sathiya, P. 2021. Methods and materials for additive manufacturing: A critical review on advancements and challenges. *Thin-Walled Structures*, 159: 107228.

Liu, K., Chen, W., Yang, W., Jiao, Z., Yu, Y. et al. 2022. Review of the research progress in soft robots. *Applied Sciences*, 13(1): 120.

Majidi, C. 2014. Soft robotics: A perspective—current trends and prospects for the future. *Soft Robotics*, 1(1): 5–11.

Majidi, C. 2019. Soft-matter engineering for soft robotics. *Advanced Materials Technologies*, 4(2): 1800477.

Mäntylä, M., Nau, D. and Shah, J. 1996. Challenges in feature-based manufacturing research. *Communications of the ACM*, 39(2): 77–85.

McCoul, David, Samuel Rosset, Samuel Schlatter and Herbert Shea. 2017. Inkjet 3D printing of UV and thermal cure silicone elastomers for dielectric elastomer actuators. *Smart Materials and Structures*, 26(12): 125022.

Miriyev, A., Stack, K. and Lipson, H. 2017. Soft material for soft actuators. *Nature Communications*, 8(1): 596.

Nof, S.Y. (ed.). 1999. *Handbook of Industrial Robotics*. John Wiley & Sons.

Peele, Bryan N., Thomas J. Wallin, Huichan Zhao and Robert F. Shepherd. 2015. 3D printing antagonistic systems of artificial muscle using projection stereolithography. *Bioinspiration & Biomimetics*, 10(5): 055003.

Rostamabadi, H., Assadpour, E., Tabarestani, H.S., Falsafi, S.R., Jafari, S.M. et al. 2020. Electrospinning approach for nanoencapsulation of bioactive compounds; recent advances and innovations. *Trends in Food Science & Technology*, 100: 190–209.

SachyaniKeneth, E., Kamyshny, A., Totaro, M., Beccai, L., Magdassi, S. et al. 2021. 3D printing materials for soft robotics. *Advanced Materials*, 33(19): 2003387.

Shaplov, Alexander S., Rebeca Marcilla and David Mecerreyes. 2015. Recent advances in innovative polymer electrolytes based on poly (ionic liquid)s. *Electrochimica Acta*, 175: 18–34.

Shen, Z., Chen, F., Zhu, X., Yong, K.T., Gu, G. et al. 2020. Stimuli-responsive functional materials for soft robotics. *Journal of Materials Chemistry B*, 8(39): 8972–8991.

Slack, N. 1983. Flexibility as a manufacturing objective. *International Journal of Operations & Production Management*, 3(3): 4–13.

Tan, X., Yu, C., Ren, Y., Cui, S., Li, W. et al. 2021. Recent advances in innovative strategies for the CO_2 electroreduction reaction. *Energy & Environmental Science*, 14(2): 765–780.

Truby, R.L. 2021. Designing soft robots as robotic materials. *Accounts of Materials Research*, 2(10): 854–857.

Wang, J. and Chortos, A. 2022. Control strategies for soft robot systems. *Advanced Intelligent Systems*, 4(5): 2100165.

Wang, L., Nurzaman, S.G. and Iida, F. 2017. Soft-material robotics. *Foundations and Trends® in Robotics*, 5(3): 191–259.

Wang, L., Yang, Y., Chen, Y., Majidi, C., Iida, F. et al. 2018. Controllable and reversible tuning of material rigidity for robot applications. *Materials Today*, 21(5): 563–576.

Yang, Y., Wu, Y., Li, C., Yang, X., Chen, W. et al. 2020. Flexible actuators for soft robotics. *Advanced Intelligent Systems*, 2(1): 1900077.

Yap, Y.L., Sing, S.L. and Yeong, W.Y. 2020. A review of 3D printing processes and materials for soft robotics. *Rapid Prototyping Journal*, 26(8): 1345–1361.

Yogeswaran, N., Dang, W., Navaraj, W.T., Shakthivel, D., Khan, S. et al. 2015. New materials and advances in making electronic skin for interactive robots. *Advanced Robotics*, 29(21): 1359–1373.

Zhang, Y. and Lu, M. 2020. A review of recent advancements in soft and flexible robots for medical applications. *The International Journal of Medical Robotics and Computer Assisted Surgery*, 16(3): e2096.

Zhao, Wenchuan, Yu Zhang and Ning Wang. 2021. Soft robotics: Research, challenges, and prospects. *Journal of Robotics and Mechatronics*, 33(1): 45–68.

Zhu, M., Biswas, S., Dinulescu, S.I., Kastor, N., Hawkes, E.W. et al. 2022. Soft, wearable robotics and haptics: Technologies, trends, and emerging applications. *Proceedings of the IEEE*, 110(2): 246–272.

CHAPTER

11 Advancing Circular Economy Practices in Manufacturing

Innovations and Applications in Sustainable Production

Amit Kumar,[1] *Rahul Singh*,[2] *Sandeep Sagar*,[3] *Deepansh Kumar*,[3] *Rohit Kumar*[3] and *Dipti Bharti*[3,*]

The circular economy is a transformative model aimed at achieving sustainable growth by maximizing resource efficiency, minimizing raw material use, and recovering waste through recycling and reuse. Emerging in response to the inefficiencies of linear economic systems, it promotes the principles of reduce, reuse, and recycle to extend product life cycles and reduce environmental impact. Inspired by nature's closed-loop systems, this approach turns waste into valuable inputs, fostering a balance between sustainability and development. As highlighted in the chapter, collaborative efforts are key to overcoming implementation barriers and establishing the circular economy as a viable path for industrial transformation.

[1] Indian Institute of Technology, Ropar, India. https://orcid.org/0009-0007-4346-9537
[2] School of Basic and Applied Sciences, IILM University, Greater Noida, India. https://orcid.org/0009-0003-2058-0081
[3] Darbhanga College of Engineering, Department of Science Technology and Technical Education, Government of Bihar, Darbhanga, India.
https://orcid.org/0009-0007-5445-8980, https://orcid.org/0009-0007-1146-4328
https://orcid.org/0009-0008-9556-5376, https://orcid.org/0000-0001-5658-2749
* Corresponding author: diptidce@gmail.com

DOI: 10.1201/9781003607359-11

11.1 Introduction

The manufacturing sector ranks as the third-largest source of global environmental pollution, responsible for over 20% of total greenhouse gas emissions. Beyond emissions, its socioeconomic impacts are globally recognized. According to the UN Environment Programme, both production and consumption in industrial cycles significantly harm the environment through heavy use of non-renewable resources and emissions of toxic gases and acids (Blomsma, 2017). Indirect environmental harm stems not only from direct emissions—such as those caused by chemical reactions and equipment leaks—but also from the energy demands of factory operations and the waste generated by mass consumption. Manufacturing processes and industrial supply chains thus remain major contributors to degradation, underscoring the urgent need for targeted industrial and policy reforms (Bocken, 2016). The circular economy offers a vital shift toward sustainable industrial development by maximizing resource use and minimizing waste. As global and regional challenges—like those in Ukraine—intensify, adopting circular strategies becomes increasingly essential for long-term ecological and economic resilience.

This chapter aims to examine both theoretical and practical foundations of sustainable manufacturing, expanding beyond the traditional focus on energy efficiency and profit to encompass economic equity, social responsibility, and corporate ethics. It seeks to guide industry leaders, policymakers, and stakeholders by highlighting the broader dimensions of sustainable practices, supported by recent research and case studies from leading firms. This foundation provides a basis for future innovation and collaborative exploration in both academic and industrial contexts. Based on corporate reports and official digital statements, the reviewed cases highlight recent, innovative sustainability efforts by global leaders across sectors such as retail, textiles, biotech, food and beverage, automotive, and IT. A systematic review, guided by the framework of sustainable manufacturing for a circular economy, was used to assess these initiatives (Fig. 11.1; Cramer, 2020). The chapter highlights evolving circular manufacturing systems and their potential to enhance resilience, reduce environmental harm, and support sustainable development. Through case studies, literature reviews, and expert interviews, it explores key strategies such as durable product design, closed-loop systems, advanced recycling, and renewable energy integration—aimed at shifting traditional manufacturing toward sustainability (Cramer, 2020).

11.2 Principles of Circular Economy in Manufacturing

The circular economy in manufacturing revolves around three core principles: reducing waste and pollution, reusing materials, and renewing natural systems.

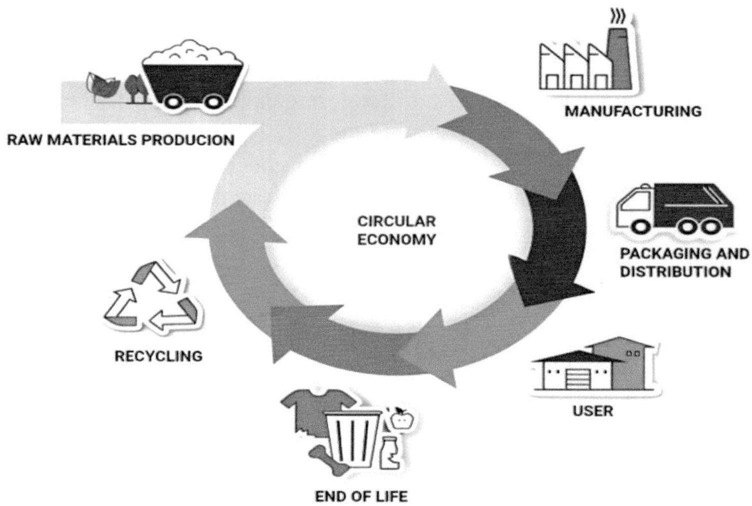

Figure 11.1: Circular economy practices in manufacturing.

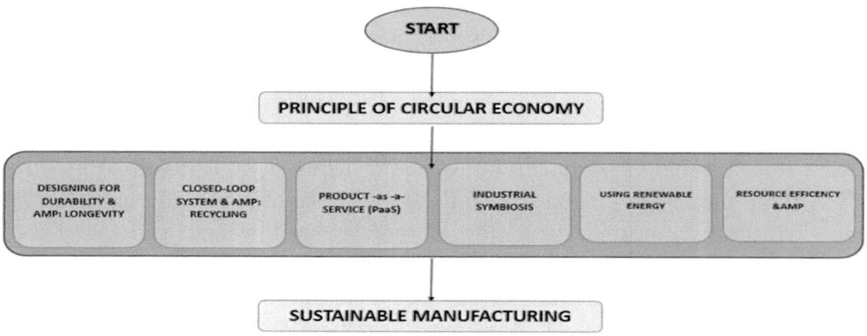

Figure 11.2: Circular economy's principles and their connection to sustainable manufacturing.

This is achieved through recycling, remanufacturing, repair, and durability-centered design (Fig. 11.2). Unlike the linear "take, make, dispose" model, the circular approach promotes extended product lifespans and efficient resource reuse. As environmental challenges escalate, these sustainable strategies are increasingly vital (Geissdoerfer, 2017).

I. Designing for Durability and Longevity

Circular economy principles emphasize designing products that are durable, easy to maintain, repair, and upgrade. Key strategies include the use of premium materials and modular designs that extend product life. For instance, Fairphone's modular smartphones allow users to replace faulty components without discarding the entire device, promoting sustainable consumption and minimizing waste (Ghisellini, 2016).

Closed-Loop Systems and Recycling

Closed-loop systems extend resource lifespans through repurposing, remanufacturing, and recycling, aiming to minimize waste and keep materials circulating within the production cycle. For example, Caterpillar Inc.'s remanufacturing program restores used components to like-new condition, significantly reducing the demand for virgin resources (Ghisellini, 2016).

Product as a Service (PaaS)

The product-as-a-service (PaaS) model shifts from ownership to access, with manufacturers retaining responsibility for maintenance throughout the product's life. This incentivizes the design of durable, easy-to-maintain goods. Rolls-Royce's "Power by the Hour" program exemplifies this approach, charging airlines based on engine usage rather than selling the engines outright (Geng, 2012).

Industrial Symbiosis

Industrial symbiosis is the collaborative use of waste and by-products across industries to boost resource efficiency and cut waste. It involves networks that share materials, energy, and information. A prime example is the Kalundborg Symbiosis in Denmark, where sectors like agriculture, power, and pharmaceuticals exchange resources to mutual benefit (Ghisellini, 2016).

Using Renewable Energy

Integrating renewable energy into manufacturing is key to advancing the circular economy and reducing environmental impact. Switching to sources like wind and solar significantly cuts carbon emissions and supports sustainable practices. Tesla's Gigafactory exemplifies this shift, operating on renewable power in line with its sustainability goals (Ghisellini, 2016).

Resource Efficiency and Optimization

Resource efficiency aims to minimize waste and environmental impact by optimizing water, energy, and material use in manufacturing. This involves adopting advanced technologies, lean practices, and continuous improvement. Toyota's Production System (TPS) exemplifies this approach, emphasizing efficiency and waste reduction. These strategies not only support eco-friendly operations but also drive innovation and growth. The following sections explore practical tactics and sector-specific case studies (Haas, 2015).

11.3 Innovations Driving Circular Economy

Innovation is central to advancing circular economy practices, driving progress across material science, technology, business models, and policies (Kirchherr, 2017; Ghisellini, 2014). As outlined in Table 11.1, breakthroughs include eco-design for durability and reuse, cleaner manufacturing processes, organizational shifts toward resource efficiency, and policy support for sustainable production. These multidisciplinary innovations are key to reshaping manufacturing for a more circular and sustainable future.

Table 11.1: Eco-innovation types for a circular economy.

Sort	Brief descriptions, examples and keywords
Eco-innovation in product design	Throughout a product's life cycle, environmental impact and material use are minimized by enabling recovery strategies like cascade use, remanufacturing, repair, maintenance, and recycling
Eco-innovation in processes	In industrial processes, improved efficiency reduces material use, emissions, hazards, and costs. Emphasis is placed on remanufacturing through repair or part replacement, along with product updates. Resource recovery is promoted via disassembly at material, component, and chemical levels, supporting functional recycling, downcycling, and upcycling—all contributing to cleaner production, zero waste, and near-zero emissions.
Eco-innovation within the organization	Management systems are being restructured to close resource loops and enhance efficiency. This includes adopting new business strategies like industrial symbiosis and developing innovative ways to recover valuable resources from products, turning them into useful services (systems of products and services).
Promoting eco-innovation	Eco-innovation in marketing, focuses on sustainable design, pricing, and promotion strategies. It encourages product reuse—both for original purposes (e.g., bottles, appliances) and new functions (e.g., tires repurposed as playground equipment). Green branding and eco-labelling further support conscious consumer choices.
Eco-social innovation	Lifestyle changes and user-driven innovation support the circular economy through collaborative consumption (e.g., shared housing, tools), product sharing (e.g., appliances, books), and sufficiency measures (e.g., plastic bag bans). These shifts encourage mindful use, smart purchasing, and prioritizing access over ownership.
Eco-innovation in systems	Developing entirely new systems with novel functionalities helps reduce environmental impact and drive industrial dematerialization. Approaches like permaculture, smart cities, and innovative urban governance support this transformation toward sustainable living.

11.4 Advances in Material Science

Advances in material science are crucial for developing recyclable, sustainable materials that enhance environmental performance and product quality. Key innovations include biodegradable polymers, recycled composites, and self-healing materials—supporting circular manufacturing goals (Homrich, 2018; Lieder, 2016).

Biodegradable Polymers: Biodegradable materials decompose naturally over time, reducing plastic waste. Companies like NatureWorks produce such plastics from renewable sources.

Recycled Composites: Enhanced recycling technologies now enable the creation of high-quality composites from reused materials, benefiting sectors like automotive and construction through improved sustainability and performance.

Self-Healing Materials: Self-healing materials are being developed to automatically repair minor damage, extending product life and minimizing waste. BASF is one of the frontrunners in this innovation.

11.5 Digital Technologies

Digital technologies play a key role in enabling circular economy practices by improving efficiency, transparency, and resource management. Core innovations include IoT for real-time monitoring, blockchain for supply chain traceability, and AI for production optimization (Su, 2013; Stahel, 2016).

Internet of Things (IoT): IoT devices enable real-time product monitoring, supporting predictive maintenance and reducing downtime. Siemens uses these solutions to enhance manufacturing efficiency.

I. **Blockchain:** Blockchain enhances supply chain transparency by tracking materials across their lifecycle, ensuring authenticity and accountability. Everledger uses it to trace the origins of high-value items like diamonds.

II. **Artificial Intelligence (AI):** AI systems optimize production, predict demand, and streamline supply chains. IBM's AI tools solutions help manufacturers cut waste and boost efficiency.

11.6 Innovative Business Models

Emerging business models prioritize sustainability and efficient resource use by minimizing waste, extending product lifespans, and promoting mindful consumption. Key models include Product-as-a-Service, the sharing economy, and circular supply chains—all driving a shift toward more responsible, low-impact production and use (Murray, 2017; Potting, 2017).

Product-as-a-Service (PaaS): Instead of selling products, companies now offer them as services—ensuring durability and simplified maintenance. Rolls-Royce's "Power by the Hour" model exemplifies this, charging customers for engine use rather than ownership.

Sharing Economy: Platforms like Airbnb and Zipcar promote shared resource use without ownership, reducing demand for new production and supporting lower overall consumption.

Circular Supply Chains: Circular supply chains aim to minimize waste by keeping materials in constant circulation. Interface demonstrates this by recycling used carpet tiles into new ones, cutting dependence on virgin resources.

11.7 Policies and Regulations Supporting Circular Economy

Government policies are vital in advancing the circular economy by encouraging sustainable practices and discouraging waste. Key approaches include Extended Producer Responsibility (EPR) and resource efficiency standards that drive eco-friendly product design and lifecycle management (Reike, 2018; Nubholz, 2017).

Extended Producer Responsibility (EPR): Extended Producer Responsibility (EPR) holds manufacturers responsible for their products' entire lifecycle, encouraging designs that prioritize durability and recyclability. The EU's Circular Economy Action Plan includes EPR measures across multiple industries.

I. **Resource Efficiency Standards:** Governments often enforce regulations to promote sustainable product design. Japan's Top Runner Program, for instance, sets energy efficiency benchmarks for appliances, driving ongoing improvement.

11.8 Collaborative Efforts for a Circular Economy

Collaboration among industry, academia, and policymakers is key to advancing the circular economy. Notable efforts include the Ellen MacArthur Foundation's initiatives, the New Plastics Economy, and CE100—all fostering knowledge exchange and scalable strategies for sustainable transformation (Mathews, 2016; McDowall, 2017).

Ellen MacArthur Foundation: This organization collaborates with businesses, researchers, and policymakers to accelerate the shift toward a circular economy.

I. **New Plastics Economy** initiative aims to redesign how plastic is used and recycled.

II. **Circular Economy 100 (CE100):** A global network that brings together companies, innovators, and cities to exchange knowledge and best practices for implementing circular economy strategies.

These innovations ranging from advanced materials and digital technologies to new business models, supportive policies, and collaborative initiatives are shaping the future of sustainable manufacturing. By embracing these changes, industries can significantly reduce environmental impact while fostering economic growth and resilience.

11.9 Case Studies of Circular Economy in Manufacturing

I. Philips: Circular Procurement and Product Design

By incorporating sustainability into its product design and procurement practices, Philips continues to lead circular economy projects (Fig. 11.3). The firm creates goods with durability and recyclability in mind, and it places a high priority on using eco-friendly materials. In addition to putting take-back procedures in place to recycle and restore discarded goods, Philips' "Circular Procurement" program makes sure that its suppliers adhere to sustainable standards. This strategy promotes ongoing innovation, reduces waste, and improves resource efficiency (Korhonen, 2018; Zhou, 2024).

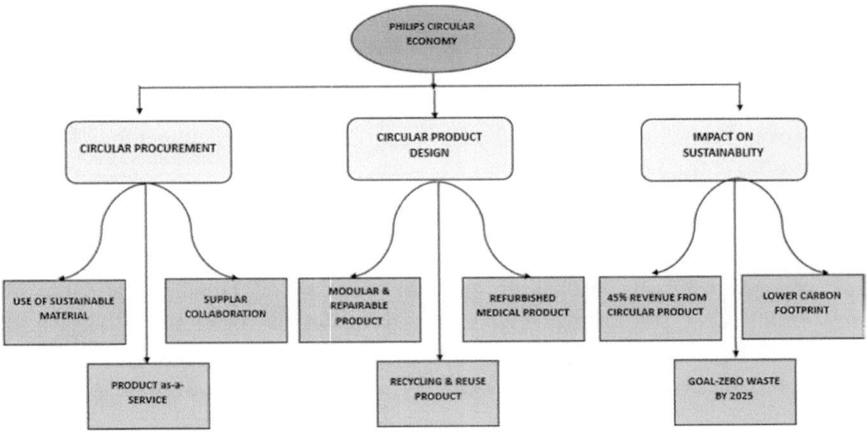

Figure 11.3: Philips' circular economy.

II. Hindustan Unilever Limited (HUL): Project Prithvi

Hindustan Unilever Limited's "Project Prithvi" showcases circular economy principles by converting collected plastic waste into fuel for cement kilns in India. In collaboration with multiple partners, HUL has processed over 5,000 metric tons of plastic, demonstrating an effective model for sustainable waste management and energy recovery (Dennison, 2024).

Indian Railways: Repurposing Discarded Tracks

Indian Railways integrates discarded tracks into its circular economy strategy, reducing the need for new material extraction and lowering pollution. This approach enhances rail infrastructure sustainability while conserving natural resources.

Godrej Group: Repurposing Wooden Pallets

Godrej Group upcycles used wooden pallets into handcrafted furniture and décor, reducing waste and emissions while supporting new business opportunities.

Goa Waste Management Corporation: Organic Waste Composting

Goa Waste Management Corporation has established composting facilities to convert organic waste into nutrient-rich compost. Used in agriculture as a natural fertilizer, this initiative reduces waste and promotes sustainable farming—demonstrating circular economy principles in action.

Reliance Industries Limited (RIL): From Fuel to Plastic Waste

Reliance Industries Limited (RIL) has set up a facility that converts plastic waste into fuel for its manufacturing power needs, reducing both plastic waste and fossil fuel use. This initiative highlights the role of circular economy strategies in waste management and sustainable energy.

LICO Materials: Battery Recycling

LICO Materials has opened a battery recycling facility in Bengaluru and plans another in Delhi-NCR, aiming to support India's growing EV sector. With efficient recycling, the company anticipates major environmental gains and targets over ₹100 crore in revenue by FY26, driven by rising demand for EVs and energy storage (de Jesus, 2018).

India-Finland Collaboration: Sustainable Packaging

Finland and India have partnered to advance eco-friendly packaging, unveiling new recyclable flexible packaging guidelines at the 2nd Think Circle event in New Delhi. Case studies presented emphasized circular economy strategies in manufacturing, demonstrating their role in waste reduction, resource conservation, and sustainable development (Saidani, 2017; Zhou, 2024).

11.10 Challenges in Implementing Circular Economy in Manufacturing

Implementing circular economy in manufacturing faces key challenges such as integrating circular practices into existing supply chains, shifting consumer behavior, transforming traditional business models, navigating unclear regulations, and managing high technology costs (Fig. 11.4).

Integration with Supply Chains: Transitioning supply chains to circular economy practices is complex, demanding strong coordination and collaboration across stakeholders.

1. **Changing Consumer Behaviour:** Shifting consumer preferences towards sustainable products and practices is essential but can be challenging to achieve.
2. **Modifying Business Models:** Transitioning from traditional linear models to circular models often requires substantial changes in business operations and strategies.

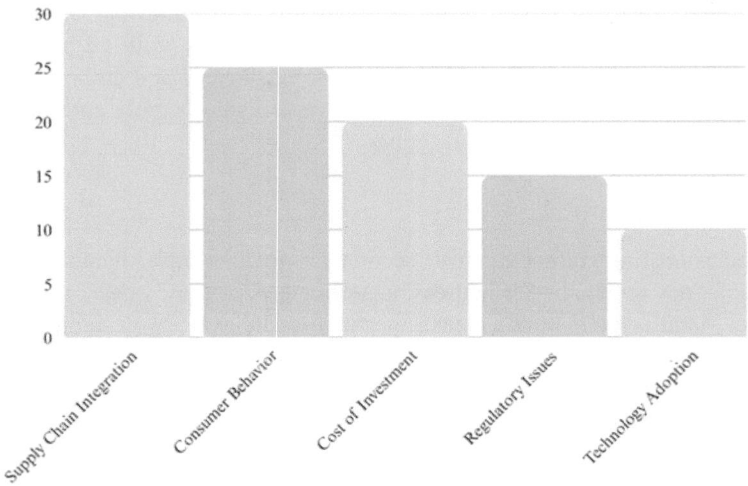

Figure 11.4: Challenges in implementing circular economy.

3. **Unclear or Lacking Regulations:** Inconsistent or inadequate regulatory frameworks can hinder the adoption of circular economy practices.
4. **Costs of New Technology Investments:** Investing in new technologies and processes for recycling, remanufacturing, and other circular practices can be expensive and may deter some manufacturers.

11.11 Future Directions of Circular Economy in Manufacturing

The core of the circular economy lies in minimizing waste, extending product lifespans, and maximizing material recovery through reuse, repair, and recycling—keeping resources in continuous circulation (Su, 2013; Stahel, 2016).

Digital Transformation: Technologies like the Internet of Things (IoT), Artificial Intelligence (AI), blockchain, and big data are likely to play a significant part in maximizing resource utilization, enhancing efficiency, and minimizing environmental footprint.

1. **Sustainable Practices:** Ongoing focus on sustainable practices, such as the employment of renewable energy, green materials, and minimization of waste, will spearhead the movement towards a circular economy.
2. **Collaborative Initiatives:** Increased collaboration between industries, academia, and governments will be essential for sharing knowledge, best practices, and resources.
3. **Supportive Legislative Frameworks:** Governments and international organizations are expected to implement policies and regulations that incentivize sustainable practices and penalize wasteful behaviours.

4. **Educational Initiatives:** Raising awareness and educating stakeholders about the benefits of circular economy practices will be crucial for widespread adoption. By addressing these challenges and embracing these future directions, the manufacturing sector can significantly contribute to a more sustainable and resilient future.

11.12 Conclusion

In conclusion, integrating circular economy principles into manufacturing is essential—not optional—for addressing resource depletion, waste, and climate change. Manufacturers have a clear opportunity to innovate by designing durable products, implementing closed-loop systems, and adopting renewable energy, reducing their environmental footprint while promoting sustainable growth.

With progress in material science, digital technology, and business models—supported by forward-looking policies and cross-sector collaboration—a solid foundation for circular practices is already in place. Despite challenges like high upfront costs and regulatory complexity, the benefits far outweigh the hurdles.

Case studies in this chapter showcase real-world success, proving the model's viability. This is a call to action: embracing circular manufacturing is vital for building a resilient economy and a sustainable future.

References

Blomosa, F. and Brennan, G. 2017. The emergence of circular economy: A new framing around prolonging resource productivity. *Journal of Industrial Ecology*, 21(3): 603–614.

Bocken, N.M.P., de Pauw, I., Bakker, C. and van der Grinten, B. 2016. Product design and business model strategies for a circular economy. *Journal of Industrial and Production Engineering*, 33(5): 308–320.

Cramer, J. 2020. The shift to a circular economy: The case of the Netherlands. *International Journal of Sustainable Development & World Ecology*, 27(6): 491–502.

De Jesus, A., Antunes, P., Santos, R. and Mendonça, S. 2018. Eco-innovation in the transition to a circular economy: An analytical literature review. *Journal of Cleaner Production*, 172: 2999–3018.

Dennison, M.S., Kumar, B. and Jebabalan, K. 2024. Realization of circular economy principles in manufacturing: Obstacles, advancements, and routes to achieve a sustainable industry transformation. *Discover Sustainability*, 5: 438.

Geissdoerfer, M., Savaget, P., Bocken, N.M.P. and Hultink, E.J. 2017. The circular economy—A new sustainability paradigm? *Journal of Cleaner Production*, 143: 757–768.

Geng, Y., Fu, J., Sarkis, J. and Xue, B. 2012. Towards a national circular economy indicator system in China: An evaluation and critical analysis. *Journal of Cleaner Production*, 23: 216–224.

Ghisellini, P., Cialani, C. and Ulgiati, S. 2016. A review on circular economy: The expected transition to a balanced interplay of environmental and economic systems. *Journal of Cleaner Production*, 114: 11–32.

Ghisellini, P., Cialani, C. and Ulgiati, S. 2014. A review on circular economy: The expected transition to a balanced interplay of environmental and economic systems. *Journal of Cleaner Production*, 114: 11–32.

Haas, W., Krausmann, F., Wiedenhofer, D. and Heinz, M. 2015. How circular is the global economy? An assessment of material flows, waste production, and recycling in the European Union and the world in 2005. *Journal of Industrial Ecology*, 19(5): 765–777.

Homrich, A.S., Galvão, G., Abadia, L.G. and Carvalho, M.M. 2018. The circular economy umbrella: Trends and gaps on integrating pathways. *Journal of Cleaner Production*, 175: 525–543.

Kirchherr, J., Reike, D. and Hekkert, M. 2017. Conceptualizing the circular economy: An analysis of 114 definitions. *Resources, Conservation and Recycling*, 127: 221–232.

Korhonen, J., Honkasalo, A. and Seppälä, J. 2018. Circular economy: The concept and its limitations. *Ecological Economics*, 143: 37–46.

Lieder, M. and Rashid, A. 2016. Towards circular economy implementation: A comprehensive review in context of manufacturing industry. *Journal of Cleaner Production*, 115: 36–51.

Marinina, O., Kirsanova, N. and Nevskaya, M. 2022. Circular economy models in industry: Developing a conceptual framework. *Energies*, 15(24): 9376.

Mathews, J.A. and Tan, H. 2016. Circular economy: Lessons from China. *Nature*, 531(7595): 440–442.

McDowall, W., Geng, Y., Huang, B., Barteková, E., Bleischwitz, R. et al. 2017. Circular economy policies in China and Europe. *Journal of Industrial Ecology*, 21(3): 651–661.

Murray, A., Skene, K. and Haynes, K. 2017. The circular economy: An interdisciplinary exploration of the concept and application in a global context. *Journal of Business Ethics*, 140(3): 369–380.

Nubholz, J.L.K. 2017. Circular business models: Defining a concept and framing an emerging research field. *Sustainability*, 9(10): 1810.

Potting, J., Hekkert, M., Worrell, E. and Hanemaaijer, A. 2017. *Circular Economy: Measuring Innovation in the Product Chain*. The Hague: PBL Netherlands Environmental Assessment Agency.

Reike, D., Vermeulen, W.J.V. and Witjes, S. 2018. The circular economy: New or refurbished as CE 3.0?—Exploring controversies in the conceptualization of the circular economy through a focus on history and resource value retention options. *Resources, Conservation and Recycling*, 135: 246–264.

Saidani, M., Yannou, B., Leroy, Y. and Cluzel, F. 2017. How to assess product performance in the circular economy? Proposed requirements for the design of a circularity measurement framework. *Recycling*, 2(6): 6.

Stahel, W.R. 2016. The circular economy. *Nature*, 531(7595): 435–438.

Su, B., Heshmati, A., Geng, Y. and Yu, X. 2013. A review of the circular economy in China: Moving from rhetoric to implementation. *Journal of Cleaner Production*, 42: 215–227.

Zhou, J., Li, X., Zhang, Z., Hou, T., Xu, J. et al. 2024. Bio-based and bio-degradable nanofiber materials: A sustainable platform for energy, environmental, and biomedical applications. *Chemical Engineering Journal*, 491(1): 152105.

Index

R

S

W

Editors Biographies

Prof. Dr. Rajesh Kumar Verma is a Professor in Mechanical Engineering at HBTU Kanpur, India, and has also served as Additional Controller of Examinations and Finance Controller. He earned his Ph.D. from Jadavpur University and is a Chartered Engineer recognized by the Institution of Engineers (India). His research spans nanomaterials, polymer composites, modeling, simulation, optimization, and advanced machining, with over 150 publications and 15 patents to his name. He has guided six Ph.D. and thirteen Master's theses, with five more Ph.D. candidates currently under supervision. Dr. Verma has led over a dozen government-funded R&D projects and has developed widely used e-content for Indian universities. He received the Research Excellence Awards in 2023 and 2024 and was ranked second in the UPPCB Hackathon "Beat Plastic Pollution." A frequent keynote speaker and reviewer for top international publishers, he is known for his contributions to sustainable material design and hybrid optimization methodologies.

Dr. Devendra Kumar Singh is a Senior Research Assistant at HBTU Kanpur, supported by UPCST Lucknow. A Life Member of the Asian Polymer Association, his research spans nanomaterial synthesis, bio-nanocomposites, numerical modeling, and surface texturing. He has published over 20 peer-reviewed papers, filed multiple design and process patents, and contributed to books and copyright-registered works. His work on biomedical nanocomposites and hybrid metaheuristic optimization has been recognized by leading international publishers.

Dr. Arpan Kumar Mondal is an Assistant Professor in Mechanical Engineering at NITTTR Kolkata. He earned his Ph.D. from IIT Guwahati in 2018, specializing in submerged arc welding, and also holds degrees from NIT Rourkela and WBUT. His research focuses on manufacturing engineering, particularly welding, composites, and numerical modeling. With over 20 journal papers, numerous conference contributions, book chapters, and patents, he brings significant innovation to the field. Active in academia since 2011, he began as a teaching assistant at IIT Guwahati before joining NITTTR.

Prof. Dr. Panagiotis Kyratsis is Head of the Department of Product and Systems Design Engineering at the University of Western Macedonia, Greece. He also leads the Institute of Traditional Architecture and the Computational Design and Digital Fabrication Lab. Holding a Ph.D. in CAD-based manufacturing simulations from the Technical University of Crete and dual M.Sc. degrees from Cranfield University (UK), his expertise spans manufacturing, CAD/CAM/CAE, reverse engineering, and product design. He has authored over 180 scientific papers, 20 books, and holds 12 industrial designs and two patents. Dr. Kyratsis also serves on editorial boards and as a reviewer for various international journals.

For Product Safety Concerns and Information please contact our EU representative GPSR@taylorandfrancis.com Taylor & Francis Verlag GmbH, Kaufingerstraße 24, 80331 München, Germany